The Blueprint of the Universe

Somewhere in the Darkness of the Cosmos, Volume 1

SANDEEP CHAVAN

Published by SANDEEP CHAVAN, 2025.

THE BLUEPRINT OF THE UNIVERSE

First edition. January 26, 2025.

ISBN: 979-8230563099

Written by SANDEEP CHAVAN.

Table of Contents

"To those who gave me life, and to the one who reshaped our understanding of it."

To My Parents, Jagannath and Baby Chavan

For your unconditional love, wisdom, and the values that shaped my relentless pursuit of knowledge. You have been my greatest inspiration, grounding me in reality while encouraging me to reach beyond the known.

To Albert Einstein

For daring to question, for redefining space, time, and gravity, and for proving that true genius lies in simplicity. Your vision laid the foundation upon which this work stands, pushing the boundaries of our understanding of the universe.

May this book serve as a bridge between generations—between those who raised me, those who inspired me, and those who will carry this knowledge forward.

"When there is nothing, there is energy."— **Sandeep Chavan**

"The universe is not a machine; it is a self-sustaining flow of energy, structured in perfect equilibrium. To understand reality, we must not seek particles alone, but the energy that binds, moves, and transforms them."

"He who sees action in inaction and inaction in action is truly wise."— **Bhagavad Gita**

This book challenges conventional wisdom, redefining gravity, motion, and existence itself—not as isolated phenomena, but as manifestations of an underlying, structured energy medium governing the cosmos.

Prologue

For centuries, science has sought to explain the fundamental forces governing the universe. From Newton's laws of motion to Einstein's relativity, and from the quantum revolution to the search for a Unified Theory, we have attempted to build a framework that explains gravity, energy, space, and time.

Yet, despite our greatest discoveries, fundamental questions remain unanswered:

- What is gravity beyond mathematical equations? Is it truly a force, or something deeper?
- Why do galaxies rotate as if invisible matter exists? What if dark matter is not real, but a symptom of an incomplete model?
- Why does space itself appear to have energy? Is the vacuum truly empty, or is it the very fabric of existence?
- What if consciousness itself is not separate from physics, but an expression of structured energy dynamics?

Modern physics has grown more complex, yet paradoxically more uncertain. Theories built upon theories, patches upon patches—yet the true nature of reality remains elusive.

A Necessary Shift in Perspective

This book does not seek to modify existing physics—it seeks to rewrite it from a new foundation. The Blueprint of the Universe presents Universal Energy Dynamics (UED), Chavan's Energy-Based Relativity (EBR), and Chavan's Law of Gravity, proposing that the universe is not governed by force-based mechanics, but by structured energy interactions.

- Gravity is not an intrinsic force, but an emergent effect of structured vacuum energy gradients.
- Motion is not a result of isolated forces, but a consequence of energy equilibrium across space.
- Black holes are not singularities, but structured energy regulators balancing cosmic interactions.
- Artificial gravity and energy-based propulsion could be the future of space travel.

What if everything we think we know about physics is just an incomplete interpretation of a deeper energy reality?

The Journey Ahead

This book is not just a theory—it is an invitation to rethink the cosmos. It offers a new model, testable predictions, and experimental directions for those willing to explore beyond conventional science.

Science is not about preserving old ideas—it is about seeking the truth, even when it means challenging everything we believe to be true.

Are you ready to step beyond Newton, beyond Einstein, beyond force-based physics?

Welcome to The Blueprint of the Universe.

Sandeep Chavan

Preface

For centuries, humanity has sought answers to the most fundamental questions of existence: *What is the universe made of? How did it begin? What governs its structure and evolution?* Despite groundbreaking advances in physics, from **Newton's laws of motion** to **Einstein's relativity** and **quantum mechanics**, a deeper mystery still remains—*what is the true nature of reality?*

This book, *The Blueprint of the Universe*, is the culmination of my relentless pursuit to answer these questions—not just within the established frameworks of physics, but through an **entirely new perspective: Universal Energy Dynamics (UED)**.

The Journey of This Book

Over the years, physicists have attempted to unify **quantum mechanics and general relativity**, explain **dark matter and dark energy**, and define **the true essence of gravity and consciousness**. Yet, key paradoxes still exist:

- Why does the universe expand in a structured manner instead of chaotic dispersion?
- Why do planets follow elliptical orbits, and why do galaxies rotate as they do—without needing exotic dark matter?
- What is gravity, if not a mere geometric warping of space-time?
- Could there be an **invisible energy medium**—the missing link that binds the cosmos?

Through deep analysis and thought experiments, I arrived at an astonishing realization:

Everything we observe in the universe is a result of structured energy interactions, maintaining equilibrium across all scales—from subatomic particles to galactic clusters.

This insight forms the **foundation of Universal Energy Dynamics (UED)**—a framework that does not contradict existing physics but instead expands upon it, fills its gaps, and proposes a new way of understanding reality.

The Core Ideas of This Book

In *The Blueprint of the Universe*, I propose that:

The universe is not just a vacuum filled with particles—it is an **active energy medium** that structures itself dynamically.

- Gravity is not an inherent force but **an emergent effect of structured vacuum energy gradients**.
- Black holes are not singularities but **cosmic energy regulators** ensuring universal equilibrium.
- Motion, planetary tilts, and galactic dynamics arise from **energy interactions, not just mass-based attraction**.

- Consciousness itself may be an **energy structuring process**, bridging the gap between physics and life sciences.

Beyond a Book—A Call for Exploration

This is not just a **theory on paper**—it is **a call for a paradigm shift**. Through **Chavan's Energy-Based Relativity (EBR) and Chavan's Law of Gravity**, I aim to inspire new experiments, practical applications in artificial gravity, space propulsion, and deeper cosmological exploration. If these ideas hold, they have the potential to redefine physics, engineering, and our understanding of existence itself.

Acknowledgment & Vision Ahead

I am deeply grateful to the thinkers, physicists, and scientists whose work laid the foundation for this book. From Newton, Einstein, and Tesla to Bose, Schrödinger, and Heisenberg, their insights inspired me to question everything. But beyond science, I am also guided by the philosophical wisdom of Indian scriptures like the Vedas and the Gita, which hint at a profound understanding of energy long before modern physics emerged.

This book is not the end—it is the beginning of a much larger exploration. If UED proves correct, it will reshape how we perceive the cosmos, gravity, time, and even life itself. And if it doesn't, it will still open doors for future thinkers to refine and expand upon these ideas.

I invite physicists, researchers, and open minds to engage with these concepts, conduct real-world experiments, and push the boundaries of what we know.

Because in the end, science is not about proving what we already believe—it is about discovering what we have yet to imagine.

Sandeep Chavan

Opening Note

We live in a time where physics has reached a paradox. On one hand, our equations successfully describe the motion of planets, the bending of space-time, and the quantum behavior of particles. On the other hand, we still don't know what gravity truly is, why galaxies rotate the way they do, or why the universe expands as if driven by an unseen force.

For over a century, we have been searching for a Unified Theory—one that can bridge the gap between General Relativity and Quantum Mechanics, explain dark matter and dark energy, and resolve the fundamental contradictions in our understanding of the cosmos.

What if the key to unification has been in front of us all along—but hidden by our assumptions?

This book, *The Blueprint of the Universe*, is an attempt to challenge our foundational understanding of physics. It proposes that:

- Gravity is not a fundamental force, but an emergent effect of structured vacuum energy.
- Motion, planetary orbits, and black holes are governed by energy equilibrium, not just mass-based attraction.
- Galaxies do not require dark matter to maintain their rotation—vacuum energy structuring explains their stability.
- Artificial gravity and space propulsion can be engineered by manipulating structured energy fields.

For centuries, we have treated space as empty and forces as fundamental—but what if the universe is, in fact, a self-sustaining energy system? What if the so-called "void" of space is an active, structured medium that governs everything—from gravity to consciousness?

This book introduces Universal Energy Dynamics (UED), Energy-Based Relativity (EBR), and Chavan's Law of Gravity, presenting a unified framework that could redefine physics as we know it.

But let me be clear: This is not the final answer—it is an invitation to rethink.

- If this theory is correct, it will reshape science forever.
- If it is incomplete, it will still bring us closer to the truth.
- Either way, the search for the real blueprint of the universe has begun.

Are you ready to question everything?

Sandeep Chavan

PART 1: Foundations of Reality

Chapter 1: Why We Need a Universal Blueprint

Humanity has always been driven by a deep curiosity to understand the universe. We have asked the same fundamental questions for centuries—what is the universe made of? How does life emerge from non-living matter? Why does everything seem to follow unbreakable laws? What is consciousness, and how does it fit into the larger reality?

Our search for answers has given rise to remarkable scientific and philosophical advancements. Physics has allowed us to understand the forces governing nature, from the motion of planets to the behavior of subatomic particles. Biology has decoded the complexity of life, from the structure of DNA to the way ecosystems function. Philosophy has helped us explore human thought, ethics, and the nature of reality itself.

Yet, despite all these advances, something feels incomplete. Our knowledge remains compartmentalized, with each field operating in isolation rather than working as part of a unified whole. We have fragmented explanations rather than a single, cohesive understanding of existence.

The Fragmented Knowledge of Science

Each major field of study offers valuable insights, but none provide the full picture.

- **Physics** explains fundamental forces like gravity, electromagnetism, and nuclear interactions, yet it struggles to reconcile the laws of quantum mechanics with general relativity. The universe behaves in one way on a large scale (where relativity governs) and another on the smallest scales (where quantum rules apply). Why? The answer remains elusive.
- **Biology** helps us understand life at the molecular level, but it rarely integrates physics beyond chemical interactions. Life is often seen as a complex system of molecules, but where does intelligence fit? What role does energy play beyond metabolism?
- **Philosophy** explores deep questions about existence, meaning, and consciousness, but it lacks a scientific framework to validate its insights. Consciousness is still regarded as an abstract phenomenon, separate from the physical universe.
- **Mathematics** provides the universal language of science, yet we are still discovering the deep connections between numbers, symmetry, and the fabric of reality.

Each of these disciplines operates within its own boundaries, but in truth, the universe does not recognize these divisions. Reality is not split into physics, biology, and philosophy—it is a single, interconnected system. What we need is a Universal Blueprint, a framework that can unify all these domains into a singular, coherent model.

Why We Need a Universal Blueprint

A universal theory must go beyond just explaining how things work—it must answer why they work the way they do. The laws of the universe must be connected, forming a self-sustaining system governed by principles of symmetry, energy balance, and transformation.

Instead of treating physics, biology, and consciousness as separate fields, we need a framework that links them together. This blueprint should explain not only how forces interact but also why everything in existence—from the smallest subatomic particles to the vast structure of galaxies—follows predictable patterns of balance and transformation.

Such a model would answer:

- How energy, matter, and information interact to create the universe we observe.
- Why the universe operates with structured laws rather than randomness.
- How life and intelligence emerge as part of a natural system rather than as anomalies.
- What role consciousness plays in the broader scheme of existence.

Nature Already Operates Within a Universal System

We are not inventing a new set of laws; we are merely recognizing and defining what nature already follows. The universe operates as a self-balancing system, where everything is interconnected.

- Ecosystems regulate themselves—predators and prey keep populations stable, plants and animals exchange oxygen and carbon dioxide in perfect cycles, and natural selection maintains adaptability.
- Physics maintains equilibrium through the conservation of energy and the symmetry of forces. Every action has a reaction, every force has a counterforce.
- Thermodynamics ensures that energy is never lost, only transformed from one form to another.
- Human-made systems also follow nature's rules. Airplanes don't break gravity; they use the forces of lift and thrust to balance against it. Our entire technological civilization functions within energy principles that nature itself follows.

No matter how advanced human progress becomes, we never operate outside the laws of nature—we are always working within them.

Beyond Unifying Forces—Unifying Reality

Our approach is not just another attempt at a "theory of everything" in physics. It goes beyond unifying forces to unifying reality itself.

- Energy is not just a quantity measured in joules or calories—it is the very essence of existence.
- Symmetry is not just a geometric concept—it is a fundamental law that governs balance in all things.
- Zero is not just a mathematical placeholder—it is the reference point of equilibrium in energy transformations.
- Consciousness is not just a byproduct of brain activity—it is an intrinsic part of universal energy dynamics.

This book is not about just one theory—it is about discovering the underlying blueprint of the universe, the invisible framework that holds everything together. It will help us answer the questions that have puzzled humanity for ages:

- What is the true nature of energy?
- Why does the universe obey unbreakable laws?
- How does everything remain in balance?
- What is intelligence, and is it merely a biological trait or a fundamental principle of the cosmos?
- What role do we, as conscious beings, play in this grand design?

Where We Go from Here

To begin, we must start with the most fundamental truth: everything is energy. Energy is not just what powers machines or fuels stars—it is the foundation of all existence. Understanding energy is the key to unifying all knowledge, from physics to consciousness, from the movement of galaxies to the birth of life.

In the next chapter, we will dive into the concept of energy, exploring why it is not just a force but the very essence of reality itself.

Chapter 2: Energy as the Fundamental Reality

The entire universe, from the smallest subatomic particle to the largest galaxy, is nothing but energy in motion. Every interaction, every transformation, and every phenomenon we observe—whether it is the fusion of stars, the movement of planets, the firing of neurons in the brain, or even human consciousness itself—is fundamentally a manifestation of energy.

For centuries, science has tried to break reality down into its simplest components. Matter was once thought to be the fundamental building block, but as we dug deeper, we discovered that matter is just energy bound into a stable form. Forces like gravity, electromagnetism, and nuclear interactions are just different ways energy expresses itself. Even thought, emotions, and intelligence are processes governed by the flow and transformation of energy.

Yet, despite its fundamental role, energy is often treated as just another physical quantity—something to be measured, conserved, or transferred. But energy is not just a part of the universe; it is the essence of everything. Understanding energy is the key to unifying all aspects of reality—physics, biology, consciousness, and even human perception.

In this chapter, we will explore why energy is not just a scientific concept but the foundational principle of existence itself. We will see how Universal Energy Dynamics (UED) provides a framework that explains not only physical forces but also the very structure of life, intelligence, and the cosmos.

2.1 The Universal Presence of Energy

Energy is the most fundamental and omnipresent reality of the universe. It is not just a force that powers machines, fuels stars, or enables biological life—it is the foundation of everything that exists. Every physical object, every thought, every force we experience is, at its core, a manifestation of energy in one form or another.

When we look at the vast cosmos, from galaxies billions of light-years away to the smallest subatomic particles, we see one common denominator—energy. Whether in motion or at rest, whether observable or hidden, everything operates within the laws of energy conservation and transformation.

Everything in the Universe is a Manifestation of Energy

For centuries, we have sought to understand the building blocks of reality. Early science assumed that matter was the most fundamental substance of the universe. The Greek philosopher Democritus proposed that everything is made of indivisible units called atoms. Later, Newtonian physics described the universe in terms of matter and motion, reinforcing the idea that physical objects were the core components of reality.

However, with the advent of modern physics, this perspective changed. In the early 20th century, Einstein's groundbreaking equation, $E = mc^2$, revealed that matter and energy are interchangeable. What we perceive as "solid" objects are, in fact, concentrated energy bound together by atomic forces.

- A single gram of mass, if fully converted into energy, would release an amount of energy comparable to the atomic bomb dropped on Hiroshima.
- Nuclear fusion in stars, the process that powers the Sun, turns matter into energy, creating light and heat that sustain life on Earth.
- Even the tiniest subatomic particles, such as quarks and electrons, are ultimately fluctuations in energy fields.

Matter, therefore, is not truly "solid"—it is energy in a structured form. From this perspective, the universe is not a collection of separate objects, but rather a vast, dynamic field of energy interactions.

Energy is Not Just a Physical Quantity—It is the Foundation of All Existence

Traditionally, energy has been treated as just another measurable quantity in physics. We calculate it in joules, track its movement in thermodynamics, and describe its properties in mechanics. But this approach underestimates the true nature of energy.

Energy is not just something that fuels motion or generates heat; it is the underlying principle that sustains all existence. Every force, every interaction, every thought, and every process—physical, biological, or psychological—can be understood as an energy transformation.

This idea extends beyond the physical sciences. In neuroscience, for example, the human brain operates through electrical signals and biochemical reactions, meaning that even thoughts, memories, and emotions are ultimately forms of energy exchange.

My book *Energy in Connection: Consciousness* Demystified supports this perspective, emphasizing that consciousness is not an independent entity but an energy-driven process. When the brain processes information, it organizes and directs energy in structured patterns, much like the physical universe operates through structured energy interactions.

Even in the context of life and death, energy remains constant. My book *Death: The Disconnection* explains how, at the moment of biological death, the energy sustaining the body does not disappear—it merely transitions. This aligns with the fundamental principle that energy can neither be created nor destroyed, only transformed.

From this perspective, life itself is not an exception to the laws of physics but a direct expression of energy at work. The forces that drive the cosmos and the mechanisms that govern living organisms are part of the same fundamental system—**Universal Energy Dynamics (UED).**

Energy Exists in Different States but Always Follows Transformation Principles

One of the defining characteristics of energy is that it never remains static—it constantly shifts from one state to another.

The First Law of Thermodynamics states that energy can neither be created nor destroyed, only converted from one form to another. This principle governs everything, from the motion of planets to the functioning of the human body.

Consider the different ways energy manifests and transforms:

- Kinetic Energy to Potential Energy – A roller coaster slows down as it reaches the top of a hill, momentarily "storing" its energy before releasing it in the descent.
- Chemical Energy to Electrical Energy – The food we eat is converted into ATP (adenosine triphosphate), which fuels cellular functions in the body.
- Nuclear Energy to Light and Heat – The Sun fuses hydrogen atoms, transforming their mass into vast amounts of radiant energy that support life on Earth.
- Thoughts as Energy Patterns – Every time we make a decision, neural signals transmit energy across synapses, creating structured patterns of activity in the brain.

No matter the form, all energy follows universal transformation principles, constantly shifting while obeying the laws of symmetry and balance.

My book Energy in Action explores this concept by showing how consciousness and biological functions are fundamentally energy processes, always in motion, always transforming. This idea perfectly aligns with the core principles of UED.

In conclusion, Energy is not just a force within the universe—it is the universe. Every physical object, every thought, every living system, and every cosmic phenomenon is an expression of energy in action. Whether structured as matter, experienced as forces, or processed in consciousness, everything ultimately follows the same fundamental energy dynamics.

Understanding energy is the key to unifying all aspects of science and philosophy. It bridges the gap between physics and consciousness, between the material and the immaterial. This realization forms the foundation of Universal Energy Dynamics (UED)—a framework that will allow us to explore the true Blueprint of the Universe.

In the next section, we will take this concept further by exploring how matter, forces, and even thought itself are not separate entities but structured manifestations of the same underlying energy.

2.2 Matter, Forces, and Energy: The Same Fundamental Essence

In the previous section, we established that everything in the universe is a manifestation of energy. This includes not only the tangible aspects of reality, such as matter and physical forces, but also the intangible ones, such as thought, intelligence, and consciousness.

Energy does not remain in a singular form—it stabilizes, transforms, and organizes itself into different states, giving rise to everything we perceive as reality. From the atomic structure of matter to the vast gravitational interactions of galaxies, from the biochemical processes of life to the neural activity behind human thought, all of these are different expressions of the same fundamental essence—energy.

Matter is Just a Condensed Form of Energy

For much of history, matter was considered the fundamental building block of the universe. It was believed that everything was composed of indivisible particles, later identified as atoms. However, the discovery of subatomic particles revealed that atoms themselves are structured energy fields.

Matter is not solid in the way it appears; it is composed of dynamic energy interactions stabilized into recognizable forms. The well-known equation $E = mc^2$ proved that mass and energy are interchangeable, meaning that what we perceive as solid objects are, in reality, concentrated energy fields.

- A tiny amount of mass can be transformed into an enormous amount of energy. This is the principle behind nuclear reactions, where splitting atoms (fission) or merging them (fusion) releases immense amounts of energy.
- In the interior of stars, nuclear fusion converts matter into energy, producing the light and heat that sustain life across planets.
- Even biological life relies on the conversion of energy—cells break down molecules to release energy, sustaining bodily functions.

On an atomic level, what we think of as "solid" matter is mostly empty space filled with energy interactions. The nucleus of an atom is a highly dense region of bound energy, while electrons exist as probabilistic energy fields surrounding it. This means that everything around us—including our own bodies—is ultimately an energy structure held together by fundamental forces.

Matter, therefore, is not fundamental—it is simply a stable configuration of energy, temporarily bound together before eventually transforming into another state.

Forces like Gravity, Electromagnetism, and Nuclear Interactions Are Energy Expressions

If matter is just a condensed form of energy, what about the forces that govern its interactions? These, too, are expressions of energy but in a more dynamic form.

Physics identifies four fundamental forces that regulate all physical interactions in the universe:

- **Gravity** – The force responsible for the attraction between masses. It governs planetary motion, galaxy formation, and even the bending of space-time.
- **Electromagnetism** – The force that controls electricity, magnetism, and light. It enables atomic interactions, chemical bonding, and the transmission of energy through waves.
- **The Strong Nuclear Force** – The force that binds protons and neutrons inside atomic nuclei, ensuring atomic stability.
- **The Weak Nuclear Force** – The force responsible for radioactive decay, playing a crucial role in energy transfer in stars.

Each of these forces is not an independent phenomenon but rather a specific way in which energy organizes itself.

- **Gravity is not a force in the traditional sense—it is the effect of energy influencing space-time.** The presence of mass-energy bends the fabric of space, causing objects to follow curved trajectories.
- **Electromagnetism is a structured interaction of energy fields.** Everything from light waves to electrical currents to chemical reactions is a result of energy fields influencing charged particles.
- **The Strong Nuclear Force is an energy mechanism that prevents atomic nuclei from disintegrating.** It is the most powerful force in nature, responsible for holding the fundamental structure of matter together.
- **The Weak Nuclear Force allows energy to transition between different states.** It enables atomic transformations and is essential to the energy production of stars.

Forces are not separate from energy—they are the structured behavior of energy acting across space and time. The universe does not require external forces to "push" or "pull" objects; rather, energy fields naturally interact to maintain balance, stability, and transformation.

Even Thought, Consciousness, and Life Processes Emerge from Energy Transitions

If matter and forces are structured expressions of energy, then what about thought, consciousness, and biological life? Are they fundamentally different, or do they also follow the same principles of energy transformation?
The answer lies in the fact that biological and cognitive processes are energy-driven interactions.

- **Every living cell functions through energy exchanges.** The food we consume is broken down into chemical energy, which powers biological processes. Without a continuous input of energy, life ceases to exist.
- **The brain operates as an energy-processing system.** Every thought, memory, and emotion is generated through the transmission of electrical and chemical signals across neural networks. These signals are not random; they follow structured energy patterns, allowing for complex cognitive functions.
- **Consciousness itself may be an emergent property of energy interactions.** Just as stars radiate energy in structured ways, the brain may be organizing energy into thought and awareness.

The same laws that govern energy in physics apply to the mind and biological systems as well. The conservation of energy states that energy cannot disappear—it can only change forms. This means that when the body ceases to function, the energy that sustained consciousness does not simply vanish; it transforms, continuing in another form.
This perspective eliminates the traditional divide between physical reality and mental existence. Instead of viewing consciousness as something separate from the universe, it can be understood as an advanced energy interaction system, governed by the same universal laws as matter and forces.
In conclusion, matter, forces, and thought are not separate aspects of reality—they are all manifestations of energy in different states.

- Matter is structured, condensed energy.
- Forces are organized patterns of energy interaction.
- Thought and consciousness are structured energy processes.

By understanding that everything in existence is an expression of energy, we move toward a unified framework that explains not only physical reality but also life and intelligence.
This perspective is central to Universal Energy Dynamics (UED), which proposes that all of reality—from the quantum level to the scale of galaxies, from biological systems to cognitive processes—is governed by energy flow and transformation.

In the next section, we will explore how the fundamental laws of energy ensure balance, stability, and continuous transformation in the universe, shaping everything from the movement of planets to the evolution of life itself.

2.3 Energy as the Unifying Factor of Reality

The realization that matter, forces, and even thought are just different manifestations of energy leads to an even deeper understanding—energy is the single, unifying factor that connects everything in existence.

Whether we examine the physical world, biological systems, or cognitive processes, energy remains the common thread that binds them together. Every interaction, transformation, and process in the universe—whether occurring in the vast cosmos or within the human brain—follows the same fundamental energy principles.

If we are to construct a true Blueprint of the Universe, it must be centered on the principle that everything exists within a dynamic energy system, where stability and transformation work in harmony to maintain universal balance.

The Common Thread Between Physics, Biology, and Consciousness is Energy Flow

Science has traditionally treated physics, biology, and consciousness as separate domains—physics studies forces and matter, biology studies life, and consciousness is often seen as an abstract phenomenon beyond scientific explanation. However, a closer look reveals that these domains are not separate at all; rather, they are different layers of the same energy continuum.

- **In physics, energy governs everything from the movement of galaxies to quantum fluctuations in empty space.** The laws of thermodynamics dictate that energy cannot be created or destroyed—only transferred or transformed.
- **In biology, life exists because of energy flow.** Every heartbeat, every neural impulse, and every chemical reaction in a living organism is driven by energy exchange.
- **In consciousness, energy manifests in the form of electrical and biochemical signals within the brain.** Thought, memory, and perception are structured patterns of energy processing.

Despite their differences in scale and complexity, all three domains are governed by the same underlying principle: energy flow. The same energy that fuels a star's nuclear fusion also drives human metabolism. The same energy principles that guide planetary motion also govern neural activity.

By understanding this fundamental connection, we see that the universe is not divided into separate systems but is a single, self-regulating energy field, manifesting in different ways.

Everything That Exists is Either a Stable or Active Form of Energy

All things—whether physical, biological, or cognitive—can be classified as either **stable or active energy states.** This duality defines how energy structures itself across different levels of existence.

Stable Energy States: These are systems in **equilibrium,** where energy is bound in a structured form. **For Example**: Atoms and molecules, where energy is contained within stable bonds. Celestial bodies like planets and stars, where energy is distributed in predictable patterns. Living organisms at rest, where metabolic energy remains regulated within biological processes.

Active Energy States: These are systems in motion and transformation, where energy is dynamically shifting between forms. **For Example:** Light waves and radiation, which transfer energy across space. Neural activity in the brain, where electric impulses constantly change states. Weather systems and ecosystems, where energy cycles through different states to maintain balance.

This concept explains why nothing in the universe is ever truly at rest—energy is always in a state of either structured stability or active transformation.

Even seemingly lifeless objects, like a rock or a mountain, contain energy locked in atomic structures. Given the right conditions, that energy can be released—whether through radioactive decay, external forces, or extreme temperatures.

Similarly, consciousness is not a fixed entity; it is a continuous process of energy flow, constantly restructuring itself based on new inputs. The mind never remains in a static state—it is always shifting between different forms of awareness, learning, and response.

Thus, everything that exists is either energy in equilibrium or energy in transition.

How Nature Continuously Redistributes Energy to Maintain Balance

The universe is not chaotic—it operates through self-regulating mechanisms that ensure energy remains balanced across all systems. Every system in nature follows a principle of energy redistribution, ensuring that excess energy never accumulates in one place for too long.

- **In cosmic systems:** Stars generate energy through nuclear fusion, releasing light and heat that radiate outward into space. When a star exhausts its fuel, it either collapses into a stable form (white dwarf, neutron star) or transitions into a more dynamic energy event (supernova, black hole).
- **In ecosystems:** Energy cycles through food chains—plants absorb solar energy, herbivores consume plants, carnivores consume herbivores, and decomposers return energy to the environment. No energy is lost; it is merely transferred between life forms.
- **In human physiology:** The body constantly regulates energy levels. When excess energy is consumed, it is stored in fat. When the body needs energy, it breaks down stored reserves. Even during sleep, the body redistributes energy to restore and heal itself.

At all levels, nature follows an energy equilibrium model, where stability and transformation coexist in a continuous cycle.

Even at the quantum level, where particles appear to pop in and out of existence, energy is not being lost or created—it is merely shifting between states. Quantum fluctuations are evidence of energy's constant movement, ensuring that the fabric of reality remains in balance.

This principle of energy redistribution explains why no system—whether physical, biological, or cognitive—can exist in complete isolation. Every part of reality depends on energy flow, constantly adapting and transforming to maintain equilibrium.

In conclusion, Energy is not just a force within the universe—it is the fabric that holds everything together.

- Physics, biology, and consciousness are not separate disciplines—they are different layers of energy interaction.
- Everything that exists is either a stable energy form (structured equilibrium) or an active energy form (in dynamic transformation).
- Nature ensures balance by continuously redistributing energy across all systems, from galaxies to neurons.

This realization fundamentally changes the way we view the universe. Instead of thinking of matter, forces, and thought as independent aspects of reality, we now see them as interconnected manifestations of the same energy field.

This concept forms the foundation of Universal Energy Dynamics (UED)—a framework that explains how all aspects of existence operate under a unified energy principle.

In the next section, we will explore the laws that govern energy, ensuring that balance, symmetry, and transformation occur in a structured, self-regulating manner across the universe.

2.4 The Laws of Energy: Conservation and Transformation

In the previous section, we established that energy is the unifying factor of reality, governing everything from physical forces to biological life and even consciousness. We also explored how energy exists in two fundamental states—stable and active—constantly shifting between forms while maintaining balance.

But how does energy maintain this balance? How does the universe ensure that energy is never lost or wasted, and why do all systems—whether cosmic, biological, or cognitive—follow predictable patterns of energy flow?

The answer lies in the fundamental laws of energy, which dictate how energy behaves, transforms, and maintains equilibrium across all systems. These laws are not just abstract principles; they are the very foundation of how reality functions.

The First Law of Thermodynamics: Energy Cannot Be Created or Destroyed

The First Law of Thermodynamics, also known as the Law of Conservation of Energy, states that:
Energy can neither be created nor destroyed; it can only be transferred or transformed from one form to another.
This is one of the most fundamental truths of the universe. No matter how energy is used, moved, or transformed, its total quantity remains constant.

- When a star burns fuel, it is not "losing" energy; rather, it is transforming nuclear energy into heat and light.
- When food is consumed, the body does not "generate" energy from nothing; it extracts chemical energy from nutrients and converts it into usable biological energy.
- Even in the death of a living organism, energy is not lost; it transitions into new forms, whether as heat, decomposition, or recycling through ecosystems.

This principle confirms that nothing in the universe is ever truly "gone"—only changed.
At the quantum level, even when a particle and its corresponding antiparticle annihilate each other, their energy does not disappear—it converts into radiation, following the conservation law.
This law extends beyond physics and applies to all forms of existence, including thought and consciousness.

- A thought does not "vanish" when forgotten; rather, it gets reshaped, stored, or replaced by new information, following the same principle of energy conservation in cognitive processes.
- Memory and learning involve the reorganization of neural energy, proving that even in the brain, energy never disappears—it only shifts between states.

Thus, the First Law of Thermodynamics is the most fundamental law of existence, ensuring that energy—whether physical, biological, or cognitive—remains constant, undergoing only transformation.

The Second Law: Energy Moves Toward Equilibrium but Remains Structured

While the First Law tells us that energy is conserved, the Second Law of Thermodynamics explains how energy flows and transforms over time.
This law states that:

In any energy exchange, the total entropy (disorder) of a system always increases, meaning that energy moves toward equilibrium but remains structured.

At first glance, this might seem like a contradiction—if entropy increases, does that mean everything eventually moves toward disorder? The answer is more nuanced.

- Energy naturally flows from higher concentration to lower concentration in an attempt to reach balance.
- Heat moves from hot objects to cooler ones, seeking thermal equilibrium.
- Electrical charges move toward areas of lower potential, seeking balance in an electric field.
- Water flows from high ground to low ground, seeking gravitational equilibrium.

However, despite this movement toward balance, nature maintains structure by channeling energy through ordered patterns.

- The formation of galaxies, solar systems, and planets follows structured energy distributions.
- The growth of plants, development of life, and functioning of ecosystems maintain structured patterns even as energy is cycled.
- The human brain, despite its chaotic neural activity, operates through structured patterns of electrical impulses, forming thoughts, memories, and consciousness.

Thus, energy does not simply "dissipate into chaos"—it follows a structured transformation process, balancing itself across all levels of existence.

Even in cosmic phenomena like black holes, where energy seems to "disappear," we now understand that black holes store information and release energy in structured ways, further supporting the Second Law.

This principle ensures that while energy is always in motion, the universe maintains a structured, self-organizing system.

Why Nature Always Seeks Balance and Efficiency in Energy Exchanges

The laws of energy transformation are not just about conservation and flow—they also explain why nature is inherently efficient.

Across all scales, from quantum physics to biological systems, energy is always distributed in the most optimal way possible.

- In physics, planets and moons follow stable orbits, minimizing energy loss while maintaining equilibrium.
- In biology, cells optimize energy usage, ensuring maximum efficiency in metabolic processes.

- In the brain, neurons fire only when necessary, conserving energy while maximizing cognitive function.

Nature never wastes energy unnecessarily—every transformation serves a purpose. This is why:

- The universe forms structured patterns like fractals, Fibonacci sequences, and naturally occurring symmetry, which are all expressions of energy optimizing itself.
- Biological evolution constantly selects for energy-efficient systems, leading to adaptations that ensure survival with minimal energy expenditure.
- The human body has self-regulating mechanisms—such as homeostasis in temperature control and automatic adjustments in metabolism—designed to optimize energy balance.

Even on a grand cosmic scale, dark energy, the mysterious force accelerating the universe's expansion, appears to be a balancing factor, ensuring that energy is evenly distributed across the cosmos.
Thus, nature follows a universal principle of energy efficiency, ensuring that all processes—whether physical, biological, or cognitive—operate within optimal balance.
In conclusion, The laws of energy transformation are the foundation of all reality.

- The First Law (Conservation of Energy) ensures that energy is never lost—only transformed.
- The Second Law (Entropy and Equilibrium) dictates that energy always moves toward balance but remains structured.
- Nature optimizes energy usage, ensuring balance and efficiency across all systems.

These laws apply universally, governing the movement of planets, the metabolism of cells, the formation of galaxies, and even the functioning of the human mind.
By understanding these laws, we gain a profound insight into how the universe self-regulates, maintains stability, and ensures the seamless flow of energy across all levels of existence.
In the next section, we will explore how symmetry plays a crucial role in maintaining this balance, ensuring that energy interactions remain ordered and structured across the universe.

2.5 The Role of Symmetry in Energy Dynamics

The laws of energy transformation ensure that energy is never lost but continuously moves toward equilibrium. However, energy does not just flow in random patterns—it follows structured, predictable rules that maintain order across the universe. One of the most fundamental principles governing this order is symmetry.

Symmetry is not just a mathematical tool or a visual aesthetic—it is a fundamental principle of nature that governs how energy interacts, transforms, and maintains balance. It ensures that the universe does not descend into randomness but instead follows structured, self-organizing patterns. From the smallest quantum particles to the vast cosmic structures, symmetry acts as nature's guiding force, ensuring that energy remains in equilibrium.

Symmetry is Not Just a Mathematical Concept—It Governs Energy Flow

In everyday life, we associate symmetry with balanced shapes, harmonious designs, and repeating patterns. But in physics, symmetry is much deeper—it determines how the fundamental forces and particles of the universe interact.

- In quantum mechanics, symmetry dictates how particles behave. The entire Standard Model of particle physics—the framework describing all known fundamental forces and particles—is based on symmetry principles.
- In relativity, symmetry defines space-time. Einstein's equations describe how matter and energy curve space, and these equations rely on symmetrical transformations that remain consistent across the universe.
- In thermodynamics, symmetry ensures that energy transitions follow structured rules. Energy flows in predictable ways, whether it is in the motion of planets, the exchange of heat, or the transmission of light.

Symmetry is the reason why the universe is ordered rather than chaotic. If symmetry did not exist, energy would disperse in unpredictable ways, leading to an unstable and incoherent universe.
Mathematically, every physical law that remains unchanged under certain transformations follows a conservation principle due to symmetry. For example:

- Conservation of Energy exists because physical laws remain unchanged over time (time symmetry).
- Conservation of Momentum exists because physical laws remain unchanged in different locations (spatial symmetry).
- Conservation of Charge exists because electromagnetic interactions obey symmetrical laws.

Thus, symmetry is not a byproduct of nature—it is the foundation that ensures the stability and predictability of energy interactions.

The Principle of Balance in Energy Interactions

Symmetry is also responsible for the universal balance observed in all energy exchanges. Every interaction in nature follows a principle of equal and opposite reactions, ensuring that energy remains in harmony.

- Newton's Third Law (Action-Reaction Pairing): Every force exerted by one object is met with an equal and opposite force. This is a direct result of energy balancing itself.
- Electromagnetic Fields: Every positive charge is naturally balanced by a negative charge, ensuring that electric fields remain stable.
- The Structure of Molecules: Chemical reactions follow precise symmetrical rules, ensuring that bonds form and break in an energy-efficient manner.

In biological systems, symmetry ensures efficient energy flow.

- The human body exhibits bilateral symmetry, optimizing movement and balance.
- DNA follows helical symmetry, ensuring that genetic information is stored and replicated efficiently.
- Neural networks are designed to process information in balanced pathways, allowing for efficient thought and decision-making.

Even in cognitive and emotional energy exchanges, symmetry plays a role. Human emotions and interactions often seek balance—whether in social relationships, decision-making, or even the pursuit of fairness. This suggests that symmetry is not just a physical concept but a fundamental principle shaping all levels of reality.

Why Opposites Exist in Nature: Positive-Negative, Matter-Antimatter, Action-Reaction

One of the most fascinating aspects of symmetry in energy dynamics is the existence of opposites in nature. Every fundamental interaction is balanced by an opposing force, ensuring that the universe remains stable.

1. Positive-Negative Charge Balance

In electromagnetism, every positive charge has a corresponding negative charge. This ensures that electrical forces remain stable, preventing energy from becoming unbalanced.

- If there were only positive charges, energy would accumulate in destructive ways, leading to an unstable system.
- The balance between protons (positive) and electrons (negative) ensures that atoms remain neutral and chemically stable.

2. Matter-Antimatter Symmetry

Physics predicts that for every particle of matter, there should be an equivalent antiparticle with the same mass but opposite charge.

- When matter and antimatter meet, they annihilate, converting completely into pure energy, following the principle of perfect energy balance.
- While the universe is composed primarily of matter, traces of antimatter exist, ensuring that energy interactions remain symmetrical.

3. Action-Reaction in Physical Interactions

Newton's Third Law states that every action has an equal and opposite reaction. This ensures that energy is never lost or accumulated unevenly—it is always redistributed in a way that maintains balance.

- A rocket moves forward by expelling exhaust gases backward, ensuring symmetrical energy transfer.
- A swimmer moves by pushing against water, generating an equal force in the opposite direction.

4. The Balance between Expansion and Contraction in the Universe

The universe itself follows symmetrical expansion patterns. While dark energy drives the expansion of the cosmos, gravitational forces counterbalance this expansion, preventing runaway acceleration.

- If only expansion existed, the universe would disperse into complete chaos.
- If only contraction existed, the universe would collapse into a singular point.
- The balance between these opposing forces ensures that galaxies form and remain stable.

This principle applies not only to physics but also to life and consciousness.

- The human mind constantly balances opposing thoughts, emotions, and decisions, ensuring structured thinking.
- Life itself exists in a balance between creation and decay, growth and entropy, activity and rest.

The existence of opposites is not a coincidence—it is a fundamental requirement for energy stability. In conclusion, Symmetry is not just a concept—it is the foundation of how energy maintains balance across all levels of existence.

- Symmetry governs energy flow, ensuring that the universe remains structured and predictable.
- The principle of balance in energy interactions explains why forces and physical interactions always remain stable.
- The existence of opposites (positive-negative, matter-antimatter, action-reaction) ensures that energy is evenly distributed and self-regulating.

Without symmetry, energy would not have a structured way to organize itself, leading to a chaotic and unstable universe. Instead, symmetry provides the guiding rules that allow energy to shape matter, create forces, and sustain consciousness.

This principle is central to Universal Energy Dynamics (UED), as it explains why energy behaves in ordered patterns and ensures equilibrium across all domains of existence.

In the next section, we will explore how Universal Energy Dynamics (UED) provides a framework that connects energy transformation, symmetry, and the fundamental nature of reality.

2.6 Universal Energy Dynamics (UED): The Framework for Everything

The principles of energy conservation, transformation, and symmetry have revealed a profound truth—energy is not just a fundamental component of the universe; it is the very essence of reality. Everything, from the smallest subatomic particles to the largest galaxies, from biological life to consciousness itself, follows the same energy-driven laws.

However, to truly unify all aspects of existence—matter, forces, life, intelligence, and the cosmos—we need a comprehensive framework that encapsulates these energy principles into one coherent model. This is where **Universal Energy Dynamics (UED)** comes in.

UED is not just a theoretical construct; it is a universal blueprint that describes how energy organizes itself, maintains stability, and drives intelligence across all domains—physical, biological, and cognitive. It is the missing framework that connects the laws of physics, the principles of life, and the emergence of intelligence into one unified system.

Introduction to UED as the Foundation of This Universal Blueprint

Universal Energy Dynamics (UED) is based on a fundamental realization:

Everything in existence is an expression of energy, following structured patterns dictated by transformation, conservation, and symmetry.

Unlike traditional physics, which studies energy in isolation within specific domains (such as mechanics, thermodynamics, or electromagnetism), UED transcends individual scientific fields and integrates them into a single system.

UED provides a multi-layered perspective on energy interactions:

- On the physical level, UED explains how energy forms stable structures (matter) and dynamic interactions (forces) while maintaining universal equilibrium.
- On the biological level, UED describes how energy sustains life, drives evolution, and enables metabolic and neural processes.
- On the cognitive level, UED explains how intelligence and consciousness emerge as structured energy flows within neural networks and beyond.

By unifying these perspectives, UED offers a complete framework that connects the physical, biological, and cognitive worlds under one energy principle.
This framework allows us to predict, explain, and understand the behavior of everything in the universe, from the laws of motion to the nature of thought.

How UED Explains Energy Transitions, Stability, and Intelligence

Energy is never static—it constantly transitions between different states while maintaining structured balance. UED explains these energy dynamics at all levels of existence.

1. Energy Transitions: The Continuous Flow of Existence

The universe is in a constant state of energy transformation. Everything that exists follows the cycle of:

- **Formation:** Energy stabilizing into matter or organized systems.
- **Transformation:** Energy shifting between states, from kinetic to potential, from mass to radiation, from thought to action.
- **Dissipation/Reorganization:** Energy moving toward equilibrium but never disappearing.

UED explains why all things—atoms, galaxies, ecosystems, and even thoughts—exist in a perpetual state of transformation.

- The fusion inside stars is a continuous process of energy transition, converting hydrogen into helium while releasing radiation.
- The life cycle of organisms follows an energy transition process, where biological energy is constantly repurposed.
- The human mind operates as an energy processor, transitioning between different cognitive states—learning, memory storage, and creative thought.

2. Energy Stability: The Mechanism Behind Universal Order

Despite continuous transformation, the universe does not collapse into chaos. This is because energy finds equilibrium through structured patterns, maintaining balance through symmetry and conservation laws.

- The orbits of planets follow stable energy distributions, ensuring gravitational equilibrium.
- The structure of atoms remains stable due to the balance of nuclear and electromagnetic forces.
- The body regulates energy balance through homeostasis, maintaining stable metabolism, temperature, and biochemical cycles.
- The brain optimizes neural energy usage, ensuring structured thought patterns without excessive energy waste.

Thus, UED explains why all systems naturally find stability despite ongoing transformations.

3. Energy and Intelligence: The Emergence of Thought and Awareness

Perhaps the most profound implication of UED is that intelligence itself is an energy-driven phenomenon.

The human brain, and possibly consciousness itself, functions as an energy processor—absorbing, organizing, and transmitting energy in structured patterns.

- Neural activity follows precise energy pathways, ensuring efficient data storage and retrieval.
- Learning and memory are not just chemical reactions—they are structured energy reorganizations within the brain.
- Creativity, intuition, and intelligence emerge from highly optimized energy flow within neural networks.

Even beyond biological intelligence, UED suggests that intelligent behavior—whether in nature, artificial intelligence, or collective consciousness—is a structured form of energy organization.

This means that consciousness is not an anomaly; it is an expected outcome of energy structuring itself efficiently. Just as galaxies form due to gravitational energy balances, thought forms due to optimized neural energy balances.

This perspective leads to profound implications:

- Could intelligence exist outside biological life, as an emergent energy structure?
- Does the universe itself exhibit intelligence through the optimization of energy across space-time?
- Can artificial intelligence reach human-like awareness if it achieves optimal energy

efficiency?

UED provides the foundation for exploring these questions, making it the most comprehensive framework for understanding not just the physical universe, but also intelligence and consciousness.

The Link between Energy, Symmetry, and the Structure of the Universe

Everything in the universe follows a structured pattern of energy organization. These patterns emerge through symmetry principles, ensuring that energy remains in equilibrium.

- The structure of atoms, molecules, and matter follows symmetrical arrangements, ensuring energy stability.
- Galaxies, planetary systems, and cosmic structures form based on symmetrical gravitational distributions.
- Biological life evolves through symmetrical genetic replication, ensuring optimized energy flow.
- Human thought follows structured symmetry—logic, pattern recognition, and balance in perception.

UED provides the unifying principle that connects energy, symmetry, and structure across all levels of existence.

- If the universe has structure, it must have an underlying energy symmetry.
- If life emerges, it must follow energy-efficient principles.
- If intelligence exists, it must be an optimized form of energy processing.

This understanding eliminates the traditional divide between physics, biology, and consciousness, proving that all of reality is governed by the same energy principles.
In conclusion, Universal Energy Dynamics (UED) is the foundation of the universal blueprint that governs all existence.

- It explains how energy transitions across different states, maintaining stability while enabling change.
- It reveals why the universe follows structured balance, ensuring equilibrium across all systems.
- It provides a framework for understanding intelligence and consciousness as structured energy dynamics.

Through UED, we can finally unify physics, biology, and cognition under one coherent principle—energy is the singular driving force behind everything in existence.

In the next section, we will explore how expanding our understanding of energy will lead to groundbreaking insights, reshaping our perception of the universe and our place within it.

2.7 Expanding Our Understanding of Energy

In the previous sections, we established Universal Energy Dynamics (UED) as the foundation of all existence. We explored how energy is not just a physical concept but the unifying force that connects matter, forces, life, and consciousness.

However, despite the fundamental role energy plays in shaping reality, modern physics tends to focus on individual forces rather than treating energy as a holistic principle. Science has made tremendous progress in understanding gravity, electromagnetism, nuclear forces, and thermodynamics, but it has not yet fully embraced energy as the single unifying reality that governs all aspects of existence.

To unlock a deeper understanding of the universe, we must expand our perspective on energy. We must move beyond the limitations of traditional physics and explore energy as the fundamental essence that drives everything—not just in physical systems, but also in biological life, intelligence, and even consciousness itself.

Why Current Physics Focuses on Forces but Overlooks Energy as a Whole

Physics has traditionally been built on the study of forces and interactions. Over the centuries, scientists have worked to explain how objects move, how particles interact, and how energy behaves under different conditions. This has led to incredible breakthroughs, including:

- **Newton's Laws of Motion and Gravity**—explaining how objects interact under the influence of force.
- **Maxwell's Electromagnetic Theory**—describing how electric and magnetic fields govern light and energy flow.
- **Einstein's Relativity**—showing that space and time are influenced by the presence of energy and mass.
- **Quantum Mechanics**—revealing that energy operates in discrete packets, giving rise to probability-based interactions at the smallest scales.

While these discoveries have given us a profound understanding of the universe, they have largely treated energy as a secondary concept—something that results from forces rather than being the primary reality itself.

Modern physics tends to view energy as:

- A derived quantity, measured in terms of work done by forces.
- A secondary effect, rather than the fundamental entity from which all forces emerge.
- A physical process, limited to mechanics, thermodynamics, and field interactions, rather than a principle that extends into life and intelligence.

This perspective has led to a compartmentalized understanding of reality, where physics explains forces, biology explains life, and neuroscience explains intelligence—without a unifying principle to connect them.

But what if energy is not just a property of the universe, but the foundation of everything? What if forces, matter, life, and thought are simply different expressions of a singular energy principle?

This is where UED offers a revolutionary shift in perspective.

The Need to Redefine Energy beyond Traditional Physics and Into Life, Intelligence, and Existence

To truly understand the universe, we must redefine energy in a way that unites all domains of knowledge—physics, biology, cognition, and beyond.

UED proposes that energy is not just a force that moves objects—it is the underlying reality that gives rise to everything, including intelligence and consciousness.

Let's consider how energy operates beyond traditional physics:

1. Energy in Biological Systems

- Every living cell functions as an energy processor, converting chemical energy into biological activity.
- Metabolism is not just a chemical reaction—it is an organized energy cycle that sustains life.
- DNA replication follows energy-efficient patterns, ensuring that life evolves in structured ways.
- Neural activity in the brain is not random but follows optimized energy pathways, ensuring thought and memory retention.

2. Energy in Intelligence and Thought

- The brain does not generate thoughts from "nothing"—it processes and restructures energy in the form of electrical and biochemical signals.
- Memory is not just stored information—it is an energy imprint shaped by past interactions and experiences.
- Creativity is an energy process, where existing knowledge is reorganized into new patterns of understanding.

This means that intelligence itself is a structured form of energy flow, optimized for decision-making, learning, and adaptation.

3. Energy in Consciousness and Perception

- Consciousness is traditionally viewed as a mysterious phenomenon beyond scientific explanation.
- However, when understood as an organized energy system, consciousness becomes an expected outcome of complex energy processing within neural networks.
- Just as galaxies self-organize based on energy balances, awareness emerges when energy organizes itself efficiently within the mind.

This leads to an even more profound question:
Is consciousness merely a biological function, or is it an intrinsic feature of energy itself?
If energy is truly universal, and if intelligence emerges as a structured form of energy processing, then consciousness may not be exclusive to biological organisms. It could be a natural property of structured energy itself, existing at multiple levels of complexity throughout the universe.
This perspective redefines how we understand intelligence—not just as a function of the human brain, but as an emergent property of energy dynamics, present in all structured systems.

Preparing to Explore the Deeper Implications of Energy in Future Chapters

By expanding our definition of energy, we open the door to profound new questions about existence:

- If energy is the foundation of all things, can we harness it in ways beyond our current understanding?
- If intelligence is an optimized form of energy processing, can artificial intelligence eventually reach a state of true awareness?
- If energy is universal and interconnected, can consciousness exist beyond individual organisms, potentially forming collective intelligence?
- If all structures in the universe follow energy-efficient principles, is the cosmos itself an intelligent system?

These questions challenge the traditional boundaries of science, philosophy, and metaphysics. They push us to rethink not only what we know about energy but also what it means to exist as part of this vast, interconnected system.
In the next chapters, we will take this exploration further by examining:

- The deeper implications of energy's role in life, intelligence, and the evolution of consciousness.

- How UED can revolutionize technology, artificial intelligence, and human understanding of reality.
- The potential for new discoveries in energy utilization, unlocking possibilities far beyond what is currently known in physics.

UED is not just an academic framework—it is a blueprint for understanding the true nature of the universe, life, and intelligence itself.

By embracing energy as the fundamental reality, we are not just redefining physics—we are redefining existence itself.

Chapter 3: The Role of Symmetry in Nature

Energy is the fundamental force that shapes all existence, constantly shifting between states while maintaining balance. But how does the universe ensure that energy remains structured rather than descending into chaos? The answer lies in **symmetry**.

Symmetry is not just a mathematical idea—it is the hidden order that governs everything, from the formation of galaxies to the structure of atoms, from the evolution of life to the way we think and behave. It ensures that energy is distributed efficiently, forces remain in balance, and systems stay stable over time.

In physics, symmetry dictates the behavior of fundamental forces, ensuring that every action has a reaction, and that physical laws remain unchanged across space and time. In biology, symmetry shapes life itself—our bodies, DNA structures, and even neural pathways follow symmetrical patterns to optimize function and survival. Even in human psychology, we instinctively seek balance—whether in emotions, decision-making, or relationships.

At its core, symmetry is nature's way of maintaining energy equilibrium. Every interaction, from atomic forces to planetary motion, from the smallest biochemical reaction to the grand flow of history, follows the principle that every action has a counterbalance.

Understanding the role of symmetry in nature is key to expanding Universal Energy Dynamics (UED), as it reveals why the universe follows structured energy patterns rather than random disorder. In this chapter, we will explore how symmetry not only shapes the physical world but also extends into life, intelligence, and human perception, ensuring that everything remains connected within the universal blueprint.

3.1 Symmetry as the Hidden Order behind Everything

The universe, despite its vast complexity, is not chaotic—it follows structured patterns that create order, stability, and predictability. At the heart of this order lies symmetry, a fundamental principle that governs everything from physical laws to biological structures and even human cognition.

Symmetry is not just a visual characteristic; it is the underlying framework that ensures energy remains balanced and systems function efficiently. Whether we examine the structure of galaxies, the formation of molecules, or the behavior of living organisms, we find that nature repeats symmetrical patterns to maintain harmony.

The reason for this is simple: symmetry provides the most efficient way for energy to structure itself. It minimizes excess energy expenditure, ensures stability, and allows systems to evolve in predictable ways. Without symmetry, the universe would be disordered, unstable, and incapable of sustaining life. Let's explore how symmetry reveals itself across physics, biology, and human behavior, shaping the very foundation of existence.

The Fundamental Presence of Symmetry in Physics, Biology, and Human Behavior

Symmetry in Physics: The Foundation of Universal Laws

Physics is built on symmetry—every fundamental force and interaction follows symmetrical principles. The stability of matter, the consistency of natural laws, and even the formation of the cosmos all depend on energy structuring itself in symmetrical ways.

- **Atomic Symmetry:** Atoms, the building blocks of matter, are arranged in highly symmetrical structures to maintain stability. Electrons orbit in predictable patterns, and molecules form through geometrically balanced bonds.
- **Fundamental Forces:** The laws of physics—gravity, electromagnetism, and nuclear forces—operate symmetrically, ensuring energy interactions remain consistent across time and space.
- **Conservation Laws:** Energy, momentum, and charge are always conserved due to symmetry principles. **For exampl**e, Newton's Third Law states that every action has an equal and opposite reaction, reinforcing energy balance.
- **Cosmic Symmetry:** The universe expands in a symmetrical pattern, forming galaxies, planetary systems, and cosmic structures that obey gravitational equilibrium.

Without symmetry, the laws of physics would not be constant, and the universe would be an unpredictable, chaotic space. Symmetry ensures order, structure, and sustainability across all energy interactions.

Symmetry in Biology: The Blueprint of Life

In the biological world, symmetry plays a crucial role in the design, function, and evolution of living organisms. Nature consistently favors symmetrical structures because they optimize energy distribution, enhance survival, and maintain balance.

- **DNA Structure:** The double-helix of DNA follows a perfectly symmetrical pattern, ensuring genetic information is copied with precision. This stability allows for accurate replication and evolution over time.
- **Anatomical Symmetry:** Most living organisms exhibit bilateral symmetry—from the

human body to animals and insects. This symmetry ensures balanced movement, efficient function, and structural stability.

- **Neural Symmetry:** The brain is designed with symmetrical hemispheres, allowing for optimized processing of information and balanced cognitive functions.
- **Biochemical Pathways:** Metabolic cycles, like the Krebs cycle, follow structured symmetrical patterns, ensuring efficient energy conversion within cells.

The reason life forms evolved with symmetry is simple—it is the most energy-efficient way to sustain biological processes. Organisms that are asymmetrical often experience functional disadvantages, leading to instability and inefficiency.

Thus, symmetry is not just a feature of life; it is the reason life can sustain itself.

Symmetry in Human Behavior: The Balance of Thought and Action

Beyond physics and biology, symmetry extends into human cognition, emotions, and behavior. Our minds are wired to seek, recognize, and create symmetry in our surroundings because it represents stability, predictability, and harmony.

- **Perception of Beauty:** The human brain naturally prefers symmetrical faces, structures, and patterns because they represent balance and health. This preference is deeply embedded in evolutionary psychology.
- **Decision-Making and Emotional Balance:** Our thought processes often follow symmetrical reasoning—we balance risks and rewards, analyze pros and cons, and seek equilibrium in choices.
- **Social Behavior and Relationships:** Human interactions often mirror symmetrical exchanges—give-and-take dynamics, mutual respect, and fairness are all examples of symmetry in behavior.
- **Mathematical Thinking:** Mathematics is based on symmetrical relationships—equations, proportions, and ratios all reflect a fundamental order that exists in nature.

Symmetry governs not only how we see the world but also how we interact with it. It shapes our perceptions, emotions, and even our understanding of reality.

Why the Universe is Not Chaotic—How Symmetry Creates Patterns, Structures, and Predictability

One of the greatest mysteries of existence is why the universe appears structured rather than random. The answer lies in symmetry as the organizing principle of energy.

- Without symmetry, energy would disperse in unpredictable ways, leading to disorder.

Instead, it follows structured patterns, ensuring that galaxies, atoms, and life evolve in stable ways.
- The laws of physics remain unchanged across time and space because of symmetrical conservation principles. If symmetry were broken, these laws would fluctuate unpredictably, making reality unstable.
- Biological evolution follows symmetrical patterns to optimize energy efficiency. Organisms develop in ways that balance function and survival, ensuring that energy is not wasted.
- Human cognition depends on symmetry for logical reasoning, problem-solving, and understanding the world.

Symmetry is the silent architect behind the universe. It ensures that energy flows efficiently, maintains order, and sustains reality across all levels of existence.

The Role of Mathematical Symmetry in Nature—Fractals, Golden Ratios, and Geometric Structures

Mathematics reveals the deep symmetry embedded in nature. Patterns such as fractals, the golden ratio, and geometric formations are all manifestations of how energy structures itself for maximum efficiency.

- **Fractals:** Many natural formations, such as snowflakes, river networks, and tree branches, follow fractal symmetry. These patterns repeat at every scale, ensuring energy is distributed efficiently.
- **The Golden Ratio:** Found in shell spirals, flower petals, and even human proportions, this ratio (approximately 1.618) appears consistently in biological and physical structures, optimizing balance and function.
- **Sacred Geometry:** Ancient civilizations recognized that geometric symmetry governs the structure of the universe. Concepts like the Fibonacci sequence, Platonic solids, and hexagonal patterns appear repeatedly in nature.

These mathematical principles confirm that symmetry is not an accident—it is the blueprint through which energy shapes reality.

In conclusion, Symmetry is the hidden order that governs physics, biology, and human behavior.

- In physics, symmetry maintains stability in forces, motion, and fundamental laws.
- In biology, symmetry optimizes function, evolution, and energy efficiency.
- In human behavior, symmetry influences perception, reasoning, and social interactions.
- In mathematics, symmetry appears in fractals, golden ratios, and universal geometric

structures.

Without symmetry, the universe would be chaotic and unstable. But because of it, energy remains structured, balanced, and capable of sustaining existence.

This leads us to a deeper understanding—symmetry is not just a visual property; it is a fundamental form of neutralization and balance. In the next section, we will explore how symmetry ensures that every force, action, and transformation in the universe has a counterbalance, maintaining equilibrium at all scales.

3.2 The Role of Symmetry in Physics

Physics is the language through which we understand the structure of the universe, and at its core, symmetry governs every fundamental force, law, and interaction. The very existence of stable matter, the predictability of motion, and the conservation of energy all depend on symmetrical balance. Without symmetry, the universe would be chaotic—laws would be inconsistent, and energy would not be able to sustain structured existence.

From the forces that bind atoms together to the gravitational pull that governs galaxies, symmetry acts as the stabilizing mechanism that ensures energy remains structured and interactions remain predictable. This principle is so fundamental that modern physics, including quantum mechanics, relativity, and the Standard Model of particle physics, is entirely built upon it.

Symmetry in Fundamental Forces—Electromagnetism, Gravity, and Nuclear Interactions

The universe is held together by **four fundamental forces**:

1. **Gravity** – Governs the large-scale structure of the universe.
2. **Electromagnetism** – Governs interactions between charged particles.
3. **The Strong Nuclear Force** – Holds atomic nuclei together.
4. **The Weak Nuclear Force** – Governs radioactive decay and energy transitions.

Each of these forces **follows specific symmetry principles**, ensuring that energy remains balanced across all interactions.

1. Gravity: Symmetry in Space and Time

Gravity is the force that governs planetary motion, galaxy formation, and the curvature of space-time. According to Einstein's General Theory of Relativity, gravity is not just a force—it is a result of energy shaping the fabric of space and time.

- Symmetry in gravity ensures that space-time behaves the same way in all directions. This means that no matter where we are in the universe, gravity follows the same rules.

- The inverse-square law of gravity follows radial symmetry—gravitational pull weakens with distance but remains evenly distributed.
- Black holes follow spherical symmetry, ensuring that their gravitational fields remain stable despite their immense energy concentrations.

Gravity's symmetry guarantees consistency and predictability, allowing planetary systems and galaxies to remain in balance.

2. Electromagnetism: The Perfect Balance of Positive and Negative

Electromagnetism governs all interactions between charged particles, from atomic bonding to the propagation of light. It follows an elegant symmetry between positive and negative charges.

- Every positive charge (proton) has a corresponding negative charge (electron), maintaining energy neutrality.
- Maxwell's Equations, which describe how electric and magnetic fields interact, follow perfect rotational symmetry, ensuring energy remains in equilibrium.
- Light waves, which are oscillations of electromagnetic fields, propagate in a symmetrical manner, maintaining energy stability across space.

Electromagnetic symmetry is what allows for the formation of stable atoms, molecules, and chemical reactions, making life possible.

3. The Strong Nuclear Force: The Symmetry That Holds Matter Together

The strong nuclear force is what binds protons and neutrons together inside atomic nuclei. Without this force, atoms would not exist.

- This force follows SU(3) symmetry, a mathematical principle ensuring that quarks (the fundamental particles inside protons and neutrons) interact in a balanced way.
- The arrangement of quarks in protons and neutrons is highly symmetrical, ensuring atomic stability.
- Nuclear reactions, such as fusion in the Sun, occur in perfectly symmetrical energy exchanges, ensuring efficient energy release.

The strong force is a perfect example of how symmetry is required to keep matter from collapsing into instability.

4. The Weak Nuclear Force: Symmetry in Energy Transitions

The weak nuclear force governs radioactive decay and particle transformations—processes that are critical for the evolution of the universe.

- This force follows a different kind of symmetry, known as CP symmetry (Charge-Parity symmetry), which ensures that particles and their mirror-image counterparts interact predictably.
- The weak force allows atoms to undergo transformations, converting neutrons into protons and vice versa, maintaining energy balance in the universe.
- It plays a key role in nuclear reactions inside stars, ensuring that energy is released in a controlled, symmetrical manner.

Although this force is weak compared to the others, its role in maintaining symmetry in energy transitions is crucial for the structure of matter.

Conservation Laws and Symmetry—How the Stability of Energy Follows Symmetry Principles

Symmetry is directly responsible for the conservation laws that govern physics. Every conservation law—whether it involves energy, momentum, or charge—exists because of symmetry.

1. Conservation of Energy: Time Symmetry

The reason energy is never created or destroyed (only transformed) is due to symmetry in time.

- If the laws of physics remain the same at all times, then energy must remain constant across time.
- This principle ensures that no matter how energy transitions—from kinetic to potential, from chemical to electrical—it always remains in balance.

2. Conservation of Momentum: Spatial Symmetry

Momentum remains unchanged in an isolated system because of symmetry in space.

- If the laws of physics are the same in all directions, then momentum cannot be spontaneously lost or gained in one direction without an equal counterbalance.
- This principle ensures the stability of planetary orbits, atomic motion, and even light waves.

3. Conservation of Charge: Electromagnetic Symmetry

Electric charge is always conserved because of the symmetry between positive and negative charges.

- A proton's positive charge is exactly equal to an electron's negative charge.
- When particles are created or destroyed, charge symmetry ensures that the total charge remains balanced.

These conservation laws prove that symmetry is the reason the universe remains stable rather than chaotic.

The Standard Model of Particle Physics—Why All Known Forces and Particles Follow Symmetrical Patterns

The Standard Model of particle physics is the most comprehensive theory explaining the fundamental forces and particles of the universe. It is entirely based on symmetry principles.

- Quarks and Leptons (the building blocks of matter) follow symmetrical arrangements.
- Force carriers (photons, gluons, W and Z bosons) interact in symmetrical exchanges.
- The Higgs field follows symmetry principles, giving particles mass in a structured manner.

Mathematically, the Standard Model is built on group symmetries (SU(3), SU(2), and U(1)), which describe how particles interact while maintaining balance.

Even the mysterious Higgs boson, which gives mass to particles, operates based on symmetry breaking, a phenomenon that shows how energy shifts from one balanced state to another while maintaining overall equilibrium.

The Standard Model confirms that the universe is not random—everything follows structured symmetry, ensuring stability and predictability.

In conclusion, Symmetry is the fundamental reason why physics remains consistent and stable.

- All fundamental forces (gravity, electromagnetism, nuclear forces) follow symmetrical patterns, ensuring energy balance.
- Conservation laws exist because of symmetry, preserving energy, momentum, and charge across all interactions.
- The Standard Model of particle physics is entirely based on symmetry, proving that nature follows structured energy arrangements.

Without symmetry, the laws of physics would not hold, and the universe would be unpredictable. But because of symmetry, energy interactions remain structured, and existence itself is possible.

This leads us to a profound realization: if symmetry governs physics, then it must also govern biology, cognition, and even intelligence. In the next section, we will explore how symmetry shapes life itself, ensuring that biological evolution follows the same energy balance principles found in physics.

3.3 Symmetry in Biological Systems

Symmetry is not just a principle governing physics—it is deeply embedded in biological systems. From the microscopic structure of DNA to the macroscopic design of living organisms, nature consistently relies on symmetry to maintain efficiency, stability, and survival.

The reason for this is simple: symmetry optimizes energy distribution. Whether in the growth of an embryo, the movement of an organism, or the processing of information in the brain, symmetry ensures that biological functions remain balanced, energy-efficient, and structurally sound.

This principle extends beyond physical structures—it plays a critical role in evolution, neural function, and even human cognition. Organisms that exhibit symmetry tend to survive longer, function better, and process information more efficiently.

Let's explore how symmetry shapes biology, evolution, and intelligence, ensuring that life remains in equilibrium.

The Symmetrical Design of Living Organisms—from DNA Structures to Anatomical Balance

At every level of biology, symmetry serves as the foundation of organization. From the molecular scale to entire body structures, organisms exhibit symmetrical patterns that ensure proper function and development.

1. DNA: The Perfect Symmetrical Blueprint of Life

- DNA, the molecule of life, follows a helical symmetry that ensures stability in genetic storage.
- Its double-helix structure consists of two complementary strands that are perfectly symmetrical, ensuring that information is copied with high precision.
- Symmetry in DNA allows for error correction during replication, reducing mutations and ensuring the continuity of life.

DNA's structure proves that nature relies on symmetry to store, replicate, and transfer biological information efficiently.

2. Bilateral Symmetry in Organisms

- Most multicellular organisms exhibit bilateral symmetry, meaning their left and right sides mirror each other.
- This design allows for balanced movement, optimized sensory perception, and efficient energy use.
- Even in organisms that appear asymmetrical externally (such as crabs or certain fish), their internal organs and neural structures maintain symmetrical patterns for efficiency.

Bilateral symmetry is an evolutionary advantage—it allows organisms to move in a stable, predictable manner, improving their ability to hunt, escape predators, and interact with their environment.

3. Radial and Fractal Symmetry in Nature

- Many plants and marine organisms follow radial symmetry, where structures radiate from a central axis (e.g., flowers, starfish, jellyfish).
- Leaves and branching patterns follow fractal symmetry, maximizing surface area for photosynthesis and nutrient absorption.

These natural patterns show that symmetry is the preferred design for maximizing energy flow in biological systems.

How Evolution Favors Symmetry for Efficiency, Stability, and Survival

The presence of symmetry in nature is not a coincidence—it is an outcome of evolutionary selection. Organisms that display symmetry tend to be more energy-efficient, structurally stable, and functionally superior.

1. Energy Efficiency in Symmetry

- Symmetrical organisms use less energy to move compared to asymmetrical ones.
- Muscles, bones, and joints are positioned symmetrically to distribute weight evenly, reducing energy loss.
- Predator-prey dynamics favor symmetrical organisms because they move more predictably and efficiently.

For example, cheetahs have perfectly symmetrical muscle alignment, allowing them to achieve incredible speeds while maintaining stability.

2. Symmetry in Natural Selection

- Studies show that organisms with higher symmetry are perceived as healthier and stronger, leading to greater reproductive success.
- Many species select mates based on symmetrical traits, indicating genetic stability.
- The most successful species over millions of years have evolved highly symmetrical structures, proving that symmetry is a key factor in survival.

3. Symmetry in Growth and Development

- During embryonic development, cells divide symmetrically to ensure proper organ formation.

- Any disruption in symmetry during early growth can lead to developmental disorders and structural imbalances.
- Evolution continuously selects for organisms that maximize symmetry because it ensures functional integrity, adaptability, and longevity.

Neural Symmetry and Brain Function—Why Intelligence Emerges from Structured Energy Flow

Symmetry is not limited to physical structures—it plays a crucial role in neural organization, brain function, and cognitive efficiency.

1. Bilateral Symmetry in the Brain

- The human brain is divided into two hemispheres, each controlling the opposite side of the body, maintaining balance in sensory and motor functions.
- This mirrored structure allows for redundancy, ensuring that if one hemisphere is damaged, the other can compensate.
- Many higher cognitive abilities—such as reasoning, creativity, and problem-solving—rely on symmetrical interactions between brain regions.

2. Neural Pathways and Information Processing

- The brain is structured with highly symmetrical neural circuits, ensuring efficient transmission of information.
- Symmetrical arrangement of neurons allows for optimized pattern recognition, learning, and memory storage.
- Disorders like dyslexia and autism have been linked to irregularities in neural symmetry, indicating its importance in cognitive function.

3. Intelligence as an Emergent Property of Neural Symmetry

- The ability to think, analyze, and make decisions emerges from structured energy flow within the brain.
- The synchronization of neural networks ensures that different brain regions communicate efficiently.
- Even artificial intelligence models are designed to mimic neural symmetry, as it leads to better pattern recognition and decision-making.

This suggests that intelligence itself is an outcome of structured symmetry—an optimized energy flow within the brain that enables thought, awareness, and problem-solving.
In conclusion, Symmetry is the foundation of biological design, ensuring stability, efficiency, and intelligence.

- DNA follows perfect helical symmetry, ensuring precise genetic replication.
- Organisms evolve symmetrical structures to optimize movement, function, and survival.
- Neural symmetry in the brain ensures efficient cognition, allowing intelligence to emerge from structured energy flow.

Symmetry is not just a feature of biological systems—it is the reason life exists in a stable, structured form. It is the underlying principle that ensures energy remains balanced, functional, and capable of sustaining intelligence.
This realization leads us to a deeper understanding: if symmetry governs both physical forces and biological life, then it must also play a crucial role in how energy interacts with itself to maintain balance.
In the next section, we will explore how symmetry is a fundamental form of neutralization and balance in the universe, ensuring that every action has a counterbalance to maintain equilibrium.

3.4 Symmetry as a Form of Neutralization and Balance

Throughout the universe, symmetry serves as a fundamental mechanism of balance. Whether in the laws of physics, the structures of biology, or the behavior of ecosystems, nature constantly seeks neutralization and stability through symmetrical interactions.
This principle is not just a feature of existence—it is the foundation of how energy self-regulates. Every interaction in nature involves opposing yet complementary forces, ensuring that energy remains equilibrated rather than accumulating in unstable ways.
From opposite charges in electromagnetism to the delicate balance of life processes in biological systems, symmetry ensures that all transformations occur in a way that maintains order rather than chaos.
This section explores how symmetry acts as nature's balancing force, ensuring that every action has a counterbalance, every process self-regulates, and every system remains stable.

Why Nature Always Seeks Balance—The Role of Symmetry in Self-Regulation

At all levels of existence, symmetry is nature's way of ensuring stability. Energy does not accumulate indefinitely in one place—it redistributes itself in patterns that follow symmetrical principles.
This principle applies universally:

- In physics, forces interact in symmetrical ways, ensuring that momentum, charge, and energy remain conserved.
- In biology, organisms regulate internal energy levels to maintain homeostasis.
- In ecosystems, predator-prey relationships naturally balance to prevent overpopulation or extinction.
- In human cognition, emotional and psychological states strive for balance, ensuring mental stability and adaptability.

Nature always corrects imbalances by creating opposing forces, restoring symmetry where energy begins to drift toward instability. This self-regulation mechanism is what keeps planets in orbit, ecosystems thriving, and biological life sustained.
Examples of Natural Self-Regulation Through Symmetry:

- **Star formation and collapse:** A star's outward radiation pressure is balanced by gravitational inward pull. This keeps the star stable for millions of years.
- **Ecosystem balance:** If prey populations increase, predator numbers also rise, restoring balance in the food chain.
- **Body temperature regulation:** When the body heats up, it sweats to cool down; when it cools down, it shivers to warm up—both are symmetrical responses.
- **Economic systems:** Financial markets naturally adjust when demand and supply shift, seeking a balance that prevents extreme fluctuations.

In every case, symmetry is the mechanism that ensures that no single force dominates unchecked. This natural tendency toward balance is why systems remain sustainable rather than collapsing into disorder.

The Positive-Negative Duality—Charge Symmetry, Matter-Antimatter Balance, and Opposite Forces

One of the most striking examples of symmetry as balance is the existence of complementary opposites. From positive and negative charges to matter and antimatter, nature ensures that every force has a counterforce, every action has a reaction, and every interaction follows conservation principles.

1. Charge Symmetry in Electromagnetism

- All electrical interactions are governed by the balance between positive and negative charges.
- Every proton (positive) has an electron (negative), ensuring that atoms remain electrically neutral.

- The electromagnetic force follows perfect symmetry, ensuring energy is distributed in a way that prevents imbalances.

Without charge symmetry, matter itself would be unstable, leading to random energy buildup and destruction.

2. Matter-Antimatter Symmetry in Particle Physics

- Physics predicts that for every particle of matter, there exists an equivalent antiparticle with an opposite charge.
- When matter and antimatter meet, they annihilate, releasing energy while maintaining total energy balance.
- Though antimatter is rare in our observable universe, its existence ensures that matter formation was a structured process rather than a random event.

This symmetry principle suggests that even the existence of physical matter is a result of carefully maintained energy balance.

3. Opposing Forces in Motion and Stability

- Newton's Third Law states that every action has an equal and opposite reaction.
- Magnetic poles (north and south) exist in complementary pairs, ensuring stability in magnetic fields.
- Structural forces in engineering rely on equal distribution of stress, preventing collapse.

Without this positive-negative duality, forces would not be balanced, and the universe would be dominated by uncontrollable energy shifts. Instead, symmetry ensures energy remains structured and predictable.

Homeostasis and Biological Energy Balance—How Symmetry Maintains Life Processes

Life itself depends on symmetrical regulation of energy within biological systems. Organisms must constantly adjust internal conditions to maintain equilibrium, ensuring that they can survive and function effectively.

This biological self-regulation is known as homeostasis, and it is entirely governed by symmetry principles.

1. Thermal Regulation: The Balance of Heat Exchange

- The human body cools itself when overheated and warms itself when cold, following a symmetrical energy exchange process.

- Sweating and shivering are opposite responses designed to maintain an optimal internal temperature.
- This process is automatic—the body constantly adjusts to maintain thermal symmetry.

2. Blood Sugar and Metabolism: Energy Distribution in Balance

- Insulin and glucagon act as opposing forces that regulate blood sugar levels.
- If blood sugar rises, insulin lowers it. If blood sugar drops, glucagon increases it.
- This symmetrical hormonal response prevents dangerous energy spikes or crashes.

3. Neural Equilibrium: The Symmetry of Thought and Emotion

- The brain operates in symmetrical networks, ensuring balanced cognitive function.
- Excessive neural activity is counterbalanced by inhibitory signals, preventing overstimulation.
- Psychological balance is maintained through emotional symmetry—dopamine and serotonin regulate mood in complementary ways.

Without these symmetrical self-regulation mechanisms, biological systems would quickly collapse due to energy imbalances.

In conclusion, Symmetry is not just a structural feature of the universe—it is the fundamental mechanism of balance.

- Nature self-regulates through symmetry, ensuring that energy is evenly distributed rather than accumulating in unstable ways.
- The existence of positive-negative duality (charge symmetry, matter-antimatter, opposing forces) ensures that energy interactions remain structured.
- Biological systems maintain energy balance through homeostasis, preventing extremes that could disrupt life processes.

Without these balancing mechanisms, life, motion, and even matter itself would be unsustainable. But because of symmetry, the universe functions as a self-regulating system, where every force has a counterforce, every action has a reaction, and every process maintains stability through energy balance.

This leads to an even deeper realization: if energy always maintains balance through symmetry, then all systems—physical, biological, and even cognitive—must follow structured energy cycles.

In the next section, we will explore how symmetry connects energy balance across all domains, revealing why every action in the universe has a counterbalance to maintain equilibrium.

3.5 The Psychological and Behavioral Aspects of Symmetry

Symmetry is not just a physical or biological principle—it also plays a profound role in human psychology, emotions, and behavior. The human brain is wired to seek, recognize, and create symmetry, not just in visual patterns but in thoughts, emotions, and social interactions.

From aesthetic preferences to decision-making, from emotional regulation to relationships, symmetry acts as a guiding force, helping the mind achieve balance, stability, and harmony. This psychological inclination toward symmetry is not arbitrary—it is a deep-rooted cognitive mechanism that ensures efficiency, predictability, and equilibrium in mental processes.

In this section, we explore how symmetry influences perception, cognition, emotions, and behavior, shaping the way humans interpret and interact with the world.

Human Perception of Symmetry—Why Our Brains Associate Symmetry with Harmony and Beauty

One of the most striking aspects of human psychology is that we are naturally drawn to symmetry. Whether in faces, art, music, or architecture, symmetrical patterns are perceived as more aesthetic, harmonious, and pleasing.

This preference is not just a cultural construct—it is hardwired into our neural processes for practical, evolutionary reasons.

1. The Symmetry-Beauty Connection

- Studies have shown that people universally prefer symmetrical faces over asymmetrical ones, associating them with health, attractiveness, and genetic fitness.
- This is because symmetry signals stability and balance, traits that indicate good health and strong genetic composition in nature.
- This preference extends beyond faces—symmetrical objects, landscapes, and designs are rated as more visually appealing.

2. Why the Brain Processes Symmetry More Efficiently

- Symmetrical patterns require less cognitive effort to process because they follow predictable structures.
- The brain uses symmetry as a shortcut for identifying objects quickly, which is why we recognize symmetrical shapes faster than irregular ones.
- This is an evolutionary advantage—symmetry often signals safety, order, and reliability,

while asymmetry can indicate disorder or danger.

3. Symmetry in Art, Architecture, and Music

- Throughout history, human civilizations have used symmetry in temples, palaces, and sculptures, reinforcing the idea that balance equals harmony.
- Music also follows symmetrical patterns—scales, rhythms, and harmonies are structured symmetrically to create pleasing compositions.
- The use of golden ratio symmetry in famous paintings (such as Da Vinci's Vitruvian Man) shows that humans instinctively integrate symmetry into creativity.

Our preference for symmetry is not just about aesthetics—it is a cognitive strategy for making sense of the world efficiently.

The Role of Symmetry in Decision-Making, Emotions, and Social Interactions

Beyond perception, symmetry plays a crucial role in human cognition, emotions, and relationships. The human mind does not function in extremes—it constantly seeks balance in thoughts, emotions, and choices. Just as physical systems self-regulate through symmetry, the brain also regulates itself through balanced decision-making and emotional stability.

1. Decision-Making and the Need for Balanced Choices

- The brain processes decisions by weighing pros and cons, risks and rewards—a symmetrical evaluation of possible outcomes.
- People naturally seek fairness and balance in choices, preferring mutual benefit over one-sided gains.
- Cognitive biases like the "golden mean" principle (avoiding extremes and seeking middle ground) are based on symmetry in decision-making.

2. Emotional Regulation Through Symmetry

- The limbic system, which controls emotions, operates through symmetrical interactions between neurotransmitters.
- Dopamine and serotonin, two key mood regulators, function in opposite yet balanced ways—dopamine drives excitement, while serotonin stabilizes mood.
- When this symmetry is disrupted (e.g., excess dopamine or low serotonin), it leads to mood disorders, anxiety, or depression.

This means that emotional well-being is directly tied to symmetrical balance in brain chemistry.

3. Symmetry in Social Interactions and Relationships

- Social relationships function on give-and-take dynamics, where balance is key to harmony.
- Reciprocity (returning favors) is a fundamental principle of human interaction, based on symmetry in behavior.
- Imbalances in relationships—whether in personal or professional life—often lead to conflicts, resentment, or instability.

From trust to cooperation, symmetry is the foundation of social stability, ensuring fairness, mutual benefit, and equilibrium in human connections.

The Balance of Opposing Thoughts and Emotions—How the Mind Finds Stability through Symmetry

Just as the universe balances matter and antimatter, positive and negative forces, the human mind also maintains balance between opposing thoughts, emotions, and cognitive processes.
The interplay of rationality vs. emotion, intuition vs. logic, hope vs. fear is a crucial example of how symmetry governs mental processes.

1. Rational vs. Emotional Thinking

- The brain is divided into logical and emotional hemispheres—the left hemisphere governs analytical thinking, while the right governs creativity and emotions.
- Balanced decision-making requires input from both hemispheres, ensuring that choices are neither too emotional nor overly rigid.

2. Fear vs. Confidence—The Psychological Symmetry of Risk

- Human behavior is shaped by a balance between fear (caution) and confidence (action).
- Too much fear leads to paralysis in decision-making, while too much confidence leads to recklessness.
- The brain regulates these opposing forces through symmetrical neural pathways, ensuring adaptive responses to challenges.

3. Conflict Resolution and Cognitive Balance

- When faced with internal or external conflicts, the mind seeks a symmetrical resolution.
- This is why people negotiate, compromise, and find middle ground, rather than clinging to extremes.

- Even in ethics and morality, justice is based on balance, ensuring fair and proportional outcomes.

Ultimately, the human mind operates on the same symmetrical principles as nature—balancing opposites to maintain stability.
In conclusion, Symmetry is deeply ingrained in human perception, cognition, emotions, and social behavior.

- We instinctively perceive symmetry as beautiful and harmonious because it represents stability and order.
- Decision-making, emotional regulation, and relationships all function on symmetrical balance.
- The brain itself follows symmetrical patterns, ensuring stability in thoughts, behavior, and mental well-being.

This realization confirms that symmetry is not just a physical law—it is a fundamental principle of existence, shaping both the external universe and internal consciousness.
If symmetry is the reason physical forces, biological life, and human thought remain structured, then it must also be the guiding principle of how energy flows through all realms of existence.
In the next section, we will explore how symmetry connects with energy balance at the deepest level—ensuring that every force, transformation, and action has a counterbalance, forming the universal equilibrium that sustains all existence.

3.6 Linking Energy Balance and Symmetry: Every Action Has a Counterbalance

Symmetry is not just about structure—it is the mechanism that maintains balance in all energy interactions. Every transformation in nature follows a structured process where energy shifts, exchanges, and reorganizes itself while maintaining equilibrium.
This principle of balance is evident everywhere:

- In physics, forces act in equal and opposite pairs, ensuring that energy is never lost but redistributed symmetrically.
- In biology, life functions through self-regulating cycles, maintaining homeostasis and internal stability.
- In cognition, thoughts and emotions interact in a balanced manner, preventing mental extremes and maintaining rationality.

The fundamental truth behind these processes is that every action, force, or transformation has a counterbalance. Whether it is a moving object, a biochemical reaction, or an emotional response, nature ensures that no force or energy shift occurs without an equal and opposite effect.

This section explores how symmetry and energy conservation are deeply intertwined, ensuring that all interactions in the universe remain in balance.

The Connection between Energy Conservation and Symmetry—How All Interactions Are Guided by Balance

At the heart of the universe lies the law of energy conservation, which states that:

Energy can neither be created nor destroyed; it can only be transferred or transformed.

However, conservation alone is not enough to ensure stability—symmetry is what organizes these energy transformations into structured, self-regulating processes.

- In physics, this principle is seen in equal and opposite forces, conservation laws, and motion dynamics.
- In biology, it governs homeostasis, metabolic cycles, and genetic replication.
- In cognition, it ensures rational decision-making, emotional stability, and mental equilibrium.

Without symmetrical redistribution, energy could accumulate unevenly, leading to disruptions, inefficiencies, and instability. Instead, the universe follows precise balancing mechanisms to ensure that all energy exchanges remain structured.

Examples of Energy Balance Through Symmetry:

- Mechanical motion: When an object moves, an opposing force resists, maintaining equilibrium.
- Thermodynamics: Heat flows from hot to cold, distributing energy symmetrically across systems.
- Electrical circuits: Voltage, resistance, and current interact in structured energy loops, ensuring stable function.
- Biological cycles: Organisms regulate oxygen, glucose, and body temperature through self-balancing processes.

Thus, symmetry is not just an aesthetic feature of the universe—it is the reason energy remains stable and sustainable across all interactions.

Newton's Third Law as an Example of Universal Symmetry—Action-Reaction in Both Physical and Cognitive Realms

One of the clearest demonstrations of symmetry in energy interactions is Newton's Third Law of Motion, which states:
For every action, there is an equal and opposite reaction.
This principle is universal, extending beyond physics into cognitive, biological, and even social systems.

1. Physical Manifestations of Action-Reaction Symmetry

- **Rocket propulsion:** The downward expulsion of gas creates an equal and opposite upward thrust.
- **Walking and movement:** Every step applies force to the ground, which pushes back with equal force.
- **Collisions and momentum transfer:** When objects collide, they exchange energy in a way that preserves total momentum symmetrically.

Every force in nature must be counteracted to maintain balance, proving that symmetry is embedded in all motion and physical processes.

2. Symmetry in Cognitive and Emotional Processes

- Beyond physics, action-reaction symmetry also applies to human thought, emotions, and decision-making.
- Rational thought follows balance. We process pros and cons before making a choice, ensuring mental equilibrium.
- Emotions balance themselves. Fear and confidence, stress and relaxation, and love and anger are opposing emotional forces that keep the mind stable.
- Social interactions depend on reciprocal energy exchange. Relationships are built on give-and-take, and communication is structured through equal engagement and response.

This means that the mind operates under the same symmetrical laws that govern motion, force, and physical interactions.

Why Every Transformation in Nature Requires a Complementary Counterbalance—Examples in Physics, Biology, and Human Thought

In all processes of transformation, there must be a counterbalancing force to maintain equilibrium. This principle applies across physical, biological, and cognitive domains.

1. Physics: The Balance of Forces and Energy Exchanges

- Matter and Antimatter: Every particle of matter has a corresponding antiparticle, ensuring

charge balance.
- Magnetic poles: Every north pole has a south pole, maintaining electromagnetic stability.
- Orbital dynamics: Planets remain in orbit due to the counteracting forces of gravity and inertia.

Without these symmetrical counterforces, energy would disperse unpredictably, making the universe unstable.

2. Biology: Homeostasis and Self-Regulation

- Blood sugar regulation: Insulin and glucagon work in opposite yet complementary ways to maintain energy balance.
- Body temperature control: Sweating and shivering are counterbalancing processes that regulate heat.
- Neural activity: The brain balances excitatory and inhibitory signals to prevent overstimulation.

Every biological function follows a counterbalancing mechanism to ensure that no single process overwhelms the system.

3. Human Thought: The Balance of Mental and Emotional Energy

- Cognitive dissonance occurs when conflicting thoughts disrupt mental equilibrium, forcing the mind to resolve them by finding a balanced perspective.
- Emotional resilience depends on counterbalancing negative experiences with positive reinforcements, ensuring psychological stability.
- Decision-making is optimized through structured comparisons, ensuring a rational, symmetrical evaluation of risks and rewards.

These examples confirm that symmetry is not just a physical property—it is the guiding principle behind all energy interactions, biological functions, and cognitive processes.

In conclusion, Symmetry and energy balance are inseparable. The universe maintains structured energy flow by ensuring that every action has a counterbalance.

- The conservation of energy follows symmetry, preventing random accumulation or dissipation.
- Newton's Third Law demonstrates universal action-reaction symmetry, ensuring that energy interactions remain structured.
- Every transformation in nature—whether in physics, biology, or human thought—follows a

balancing mechanism to maintain equilibrium.

This realization suggests that the universe itself is structured as a vast energy system where forces, processes, and consciousness evolve through symmetrical energy exchanges.
In the next section, we will explore how this principle extends beyond interactions into the very structure of reality itself, revealing a deeper understanding of how energy and symmetry form the universal blueprint.

3.7 The Role of Symmetry in the Universal Blueprint

Throughout this chapter, we have explored how symmetry is not just a pattern—it is the fundamental principle that governs energy balance, physical laws, biological stability, and even human cognition. Every process in the universe follows structured, symmetrical interactions, ensuring that energy remains conserved, forces remain balanced, and systems maintain order.
This realization leads to a profound question:
If symmetry is the mechanism that maintains energy equilibrium across all domains, could it also be the guiding framework behind the structure of the universe itself?
This is where Universal Energy Dynamics (UED) comes in. UED proposes that energy is the fundamental reality, and symmetry is the organizing principle that ensures its structured flow. By integrating symmetry into UED, we can begin to see how the universe functions as a self-sustaining system where order and transformation are seamlessly connected.

How Symmetry Integrates into Universal Energy Dynamics (UED)

UED presents the universe as a dynamic, self-organizing system of energy flow, transformation, and balance. Symmetry is the key principle that ensures:

- Energy remains structured rather than dispersing chaotically.
- Matter and forces interact predictably, maintaining the stability of physical systems.
- Biological life evolves through optimized patterns, ensuring sustainability.
- Cognitive processes follow symmetrical balance, enabling rational thought and emotional stability.

By integrating symmetry into UED, we gain a complete framework that connects the behavior of energy across physics, biology, cognition, and even cosmic evolution.

1. Symmetry as the Foundation of Energy Transformation

- Energy does not remain static—it is constantly shifting between states.
- These transformations follow symmetrical laws, ensuring that every process has a counterbalance.

- Whether in the conservation of energy, the balance of forces, or the structure of atoms, symmetrical transitions maintain universal equilibrium.

2. Symmetry as the Organizing Principle of Physical Reality

- In physics, forces and particles follow symmetrical patterns, ensuring that interactions remain structured.
- The Standard Model of particle physics is entirely based on symmetry principles, demonstrating that fundamental particles arise from balanced interactions.
- The expansion of the universe follows structured, symmetrical laws, preventing random or chaotic growth.

3. Symmetry as the Mechanism for Biological Evolution

- Evolution is not random—it follows structured, energy-efficient pathways driven by symmetry.
- The bilateral symmetry of organisms ensures balance and efficiency, optimizing survival.
- Genetic replication follows symmetrical patterns, ensuring stability in the transmission of life.

4. Symmetry as the Guiding Force in Cognition and Intelligence

- The brain operates on symmetrical neural networks, ensuring efficient data processing and decision-making.
- Emotional regulation depends on a balance between opposing states, preventing mental instability.
- Human relationships follow reciprocal patterns, ensuring fairness and stability in social interactions.

This means that symmetry is not just a passive feature of the universe—it is the active framework that guides the structured evolution of everything.

Why Symmetry is the Missing Link between Order and Transformation in the Universe

One of the greatest mysteries in science has been the relationship between order and change.

- How does the universe maintain stability while allowing evolution and transformation?

- How do natural laws remain constant while enabling dynamic change?
- How do biological systems evolve without descending into chaos?

The answer lies in **symmetry.**
Symmetry is what allows:

- Energy to flow while remaining structured.
- Matter to transform while maintaining stability.
- Life to evolve while following predictable rules.
- Consciousness to emerge while sustaining balance.

Without symmetry, the universe would be either too rigid to allow transformation or too chaotic to maintain order. Instead, it follows structured, symmetrical processes that ensure both stability and evolution.
This insight suggests that symmetry is the fundamental bridge between what exists and how it changes.
It is the missing link that explains:

- Why energy organizes itself rather than dispersing randomly.
- Why nature follows predictable laws while remaining dynamic.
- Why intelligence arises as an extension of structured energy patterns.

By understanding symmetry as the hidden structure behind all change, we begin to see the universe not as a collection of disconnected forces, but as a singular, self-regulating system governed by energy balance.

Preparing to Explore How Symmetry Extends into Intelligence, Evolution, and the Expansion of the Cosmos

If symmetry governs the fundamental laws of physics and life, then it must also extend into the higher complexities of intelligence, evolution, and cosmic expansion.
In the coming sections, we will explore:

The Evolution of Intelligence

- How cognition and awareness emerge as structured energy processes.
- Why artificial intelligence mimics neural symmetry to enhance learning.
- How intelligence follows self-organizing patterns similar to biological evolution.

The Expansion of the Universe and Symmetry in Cosmic Evolution

- How the universe expands symmetrically rather than randomly.
- Why gravitational balance ensures the formation of galaxies and star systems.
- How black holes, dark matter, and dark energy follow symmetrical laws to maintain universal equilibrium.

The Role of Symmetry in Consciousness and Reality Perception

- How awareness itself may be an outcome of structured energy balance.
- Why the mind perceives order and patterns as a reflection of universal symmetry.
- How spiritual and philosophical concepts of balance align with scientific principles of symmetry.

By integrating symmetry into Universal Energy Dynamics (UED), we are not just describing physics—we are revealing the blueprint of reality itself.
In conclusion, Symmetry is the guiding framework that ensures energy remains structured across all domains of existence.

- It provides stability in physics, ensuring forces, matter, and energy remain in balance.
- It enables biological evolution, optimizing survival and efficiency through structured adaptation.
- It governs cognition and intelligence, ensuring rational thought, emotional balance, and decision-making.
- It explains the balance between order and transformation, revealing why the universe evolves in a structured, self-regulating manner.

By incorporating symmetry into UED, we gain a deeper understanding of why the universe is not random but follows structured patterns of balance, evolution, and intelligence.
This leads us to a profound realization:
If energy is the foundation of reality, and symmetry is the mechanism that structures it, then the universe itself is a dynamic, evolving system where everything is interconnected through energy equilibrium.
In the next section, we will expand on this concept, exploring how symmetry not only explains the present structure of the universe but also its continuous expansion, the evolution of intelligence, and the emergence of higher-order systems.

This will allow us to bridge the gap between physics, life, and intelligence, offering a truly unified understanding of existence.

Chapter 4: Zero – The Ultimate Reference Point

Throughout our exploration of symmetry and energy balance, we have seen that the universe operates on precise equilibrium principles. Every force has a counterforce, every transformation has a balancing mechanism, and every interaction follows structured conservation laws. At the heart of this balance lies a single, profound concept—Zero.

Zero is not just a number; it is the ultimate reference point for all existence. It represents neutrality, equilibrium, and the foundation upon which all transformations occur. In physics, zero is the baseline for measuring energy states, ensuring that every fluctuation, interaction, and force returns to a state of balance. In mathematics, zero is the backbone of numerical systems, defining the difference between positive and negative, presence and absence, action and rest.

This remarkable concept, gifted to the world by the great Indian mathematician and astronomer Mahashri Aryabhata and later formalized by Brahmagupta, revolutionized human understanding. It allowed civilizations to develop advanced mathematics, algebra, calculus, and even modern physics. Yet, despite its profound significance, zero remains one of the least understood concepts in science and philosophy.

In this chapter, we will redefine zero in the context of Universal Energy Dynamics (UED), showing that it is not merely a mathematical symbol but the cosmic anchor of balance. We will explore how zero represents the neutral energy state, the foundation of conservation laws, and the ultimate equilibrium between existence and non-existence.

By understanding zero as the fundamental reference point of the universe, we can bridge the gap between mathematics, physics, and reality itself, revealing why all forms of energy, motion, and intelligence originate and return to zero.

4.1 The Significance of Zero in Mathematics and Physics

Zero is one of the most profound and transformative concepts in human intellectual history. It is more than just a number—it is the foundation of all mathematical operations, the ultimate reference point in physics, and the equilibrium state that governs energy transformations. Without zero, modern science, mathematics, and physics would be impossible.

Yet, the journey of zero from a philosophical idea to a concrete mathematical concept was long and complex. Ancient civilizations struggled with the idea of 'nothingness,' often avoiding it in numerical systems. The Greeks, for example, refused to acknowledge zero as a number because they saw it as an absence rather than a definable quantity. However, it was in **India**, with the works of **Mahashri Aryabhata** and **Brahmagupta**, that zero was not only recognized but also mathematically defined, leading to a revolution in human thought.

This section explores how zero emerged, why it became essential for mathematics and physics, and why it remains the ultimate reference point for understanding reality.

How Zero Emerged as a Fundamental Concept in Human Understanding

The idea of zero did not develop overnight—it took centuries for civilizations to fully grasp its importance.

- Early counting systems (Egyptians, Babylonians, Romans) did not have a symbol for zero. Their number systems were based on additive methods (e.g., writing 10 as two fives or five twos).
- The Mayans independently developed a concept of zero around 4th century CE, using it in their calendars, but it was not integrated into broader mathematical use.
- Indian mathematicians, led by Aryabhata (5th century CE) and later Brahmagupta (7th century CE), were the first to formalize zero as a number, not just as a placeholder.

The Breakthrough of Brahmagupta's Zero (7th Century CE)

- Brahmagupta defined the mathematical operations of zero (addition, subtraction, and multiplication involving zero).
- He introduced the idea that zero is the result of subtracting a number from itself, setting the foundation for algebra.
- This breakthrough allowed for negative numbers, equations, and advanced calculations, forming the basis of modern mathematics.
- The Indian discovery of zero traveled through the Arabic world, where scholars like Al-Khwarizmi refined it, and eventually reached Europe in the 12th century, transforming global mathematics.

Without zero, modern algebra, calculus, and even physics would not have been possible. It was this understanding that allowed humanity to develop precise mathematical models to describe the universe.

The Role of Zero as a Numerical Placeholder and a Defining Reference Point

Zero's role is twofold—it acts as both a placeholder and a reference point.

1. Zero as a Placeholder: The Foundation of the Decimal System

- The invention of zero enabled the positional number system, which allowed numbers to be written efficiently.
- In the Roman numeral system (without zero), writing large numbers was complex and impractical (e.g., 1000 = M, 10,000 = $\overline{X}$).
- The Indian-Arabic system, which uses zero as a placeholder, made arithmetic operations much simpler: 507 is not the same as 57—zero gives position and meaning. Similarly, 10, 100, 1000—zero enables scaling and multiplication.

Without zero as a placeholder, modern calculations, logarithms, and computational algorithms would be impossible.

2. Zero as a Defining Reference Point in Mathematics

- Zero separates positive and negative numbers, forming the foundation of the number line.
- It serves as the origin in coordinate geometry, allowing precise mathematical modeling of physical reality.
- It defines neutral points in equations, enabling concepts like symmetry, balancing forces, and even complex algebraic structures.

In many ways, zero is not just a number—it is the reference point from which all other values are measured.

Why Modern Mathematics and Physics Cannot Exist Without Zero

Zero is more than a mathematical convenience—it is the foundation of every major breakthrough in physics and science. Without zero, our understanding of the universe would be incomplete.

1. Zero in Calculus and Mathematical Analysis

- Calculus, developed by Newton and Leibniz, relies on the concept of limits, where values approach zero to define derivatives and integrals.
- Differential equations, which describe motion, growth, and physical processes, are based on infinitesimal changes—essentially, movements toward or away from zero.
- Without zero, physics could not model velocity, acceleration, or any dynamic system.

2. Zero in Classical and Relativistic Physics

- In Newtonian mechanics, zero serves as a point of rest (inertia) and balance in motion equations.
- In Einstein's theory of relativity, zero plays a critical role in space-time symmetry and energy-

mass equivalence.
- The famous equation $E = mc^2$ suggests that mass can be converted to energy—but at zero velocity, the rest energy remains a fundamental state.

3. Zero in Quantum Mechanics and the Nature of Reality

- In quantum physics, zero-point energy is the lowest possible energy state of a system, showing that even in "nothingness," energy fluctuations exist.
- The Heisenberg uncertainty principle relies on zero as the foundation for probability calculations.
- Wave functions collapse to zero probability in certain states, defining quantum interactions.

Zero is not just a theoretical construct—it is the foundation upon which physics models reality.
Thus, Zero is the backbone of mathematics and physics, allowing for:

- Positional notation in arithmetic, leading to efficient computation.
- The development of calculus, which models change and motion.
- The foundation of coordinate systems, enabling precise mathematical modeling.
- The conservation of energy, relativity, and quantum mechanics, making sense of the physical universe.

Without zero, humanity would not have been able to develop the modern scientific framework that explains reality.
However, despite its deep mathematical and physical significance, zero is often misunderstood as "nothingness." But in reality, zero is not the absence of something—it is the state of perfect balance.
This realization allows us to redefine zero not just as a mathematical tool, but as the fundamental reference point of all existence.
In the next section, we will explore how zero represents equilibrium in energy transformations, serving as the ultimate neutral state in nature. This will reveal why all forces, movements, and fluctuations in the universe always seek to return to zero—the state of perfect balance.

4.2 Zero as the Representation of Equilibrium

Zero is more than just a number—it is the fundamental reference point of balance, neutrality, and symmetry in nature. It serves as the universal anchor for mathematics, physics, and all measurable interactions, ensuring that systems remain stable and energy remains structured.
Despite the fragmentation of scientific theories over time, mathematics has remained consistent, sustained by the presence of zero. This is because zero is not an arbitrary human invention—it

is a direct reflection of nature's inherent symmetry. Whether in physical laws, biological systems, or cognitive processes, zero represents the neutral state where opposing forces meet, ensuring equilibrium in all transformations.

As we unknowingly develop within nature's mysteries, we continuously uncover the deeper role of zero—not just as a mathematical tool but as a fundamental principle of existence.

Zero as the Symbol of Balance, Neutrality, and Symmetry in Nature

In every domain of science, zero serves as the reference point for equilibrium. Systems in nature do not operate in extremes—they function by maintaining balance, self-regulation, and conservation of energy. Zero is the mathematical representation of this balance, defining how forces, charges, and even thoughts interact in symmetrical ways.

1. Zero as the Neutral Point in Physical Systems

- In mechanics, an object at rest has zero net force, meaning all opposing forces cancel out.
- In electromagnetism, an electrically neutral atom has zero net charge, ensuring atomic stability.
- In thermodynamics, absolute zero represents the state where molecular motion reaches perfect stillness, the lowest possible energy equilibrium.

2. Zero as the Foundation of Symmetry in Energy Interactions

- In Newtonian mechanics, every force has an equal and opposite reaction, ensuring that systems return to zero net force.
- In thermodynamic systems, heat moves from high to low temperatures, seeking zero energy imbalance.
- In electromagnetic interactions, all dipole-dipole forces exist in a state of charge balance, ensuring matter retains its form.

From planetary orbits to atomic structures, zero is the universal stabilizer that prevents systems from collapsing or spiraling into chaos.

Zero and the Laws of Conservation: The Mathematical and Physical Anchor

Zero is deeply intertwined with the fundamental laws of conservation, which govern how energy, momentum, and charge remain balanced in the universe. These conservation laws are only possible because zero provides a mathematical reference for maintaining stability in all transformations.

1. Zero and the Conservation of Energy

- The First Law of Thermodynamics states that energy cannot be created or destroyed—only transformed.

- However, in every closed system, the net total energy remains constant, ultimately balanced at zero.

2. Zero and the Conservation of Momentum

- In motion, an object's momentum before impact equals its momentum after impact, ensuring symmetry.
- The total momentum of an isolated system remains at equilibrium, meaning that energy shifts but does not accumulate infinitely in one direction.

3. Zero and the Conservation of Charge

- In atomic structures, for every positive charge, there must be an equal negative charge, ensuring net neutrality.
- This principle ensures that matter retains its stable form through electromagnetic balance.

Across all these principles, zero is the anchor that keeps natural laws structured and predictable, allowing systems to evolve without disorder.

Zero as the Stabilizing Force in Mathematics, Energy Systems, and Thought Processes

Since zero represents the neutral state of all systems, it serves as the foundation for mathematical equations, physical stability, and even cognitive balance.

1. Zero in Mathematics: The Universal Reference Point

- Algebraic equations are structured around zero balance, ensuring that mathematical expressions remain symmetrical.
- Coordinate geometry uses zero as the origin, enabling precise spatial measurements.
- Calculus relies on zero to define limits, derivatives, and infinitesimal changes in motion.

2. Zero as the Reference in Energy Systems

- In quantum mechanics, zero-point energy represents the lowest possible energy state, showing that even in "nothingness," structured fluctuations exist.
- In thermodynamics, absolute zero represents the theoretical state of no molecular motion, the ultimate energy equilibrium.

3. Zero as the Mental and Emotional Balancing Force

- In cognitive science, the human brain naturally seeks equilibrium, balancing logic and emotion.
- Emotional resilience depends on neutralizing stress with relaxation, fear with confidence, and action with contemplation.
- Even decision-making follows symmetrical reasoning, weighing risks and rewards to arrive at a stable outcome.

This means that zero is not just a number—it is the organizing principle of balance across physical, biological, and intellectual realms.

Why Mathematics Survived Despite Fragmented Scientific Theories—Because of Zero

Throughout history, scientific theories evolved, fragmented, and were restructured, yet mathematics remained stable. The reason is simple: Mathematics is built upon zero, the universal reference point that never changes.

- Classical mechanics, relativity, and quantum physics—each operates on different scales, yet all rely on zero as the foundation of conservation laws.
- Different interpretations of space-time, motion, and energy still reference zero to define initial states, boundary conditions, and equilibrium.
- Even when paradigms shift in physics or biology, mathematical models remain valid because zero ensures that all transformations remain symmetrical.

Thus, unknowingly, we are always developing within nature's deeper mysteries, following its fundamental symmetry.
In conclusion, Zero is the backbone of mathematics, the stabilizer of physics, and the foundation of energy equilibrium.

- It represents balance, neutrality, and the structured transformation of all energy interactions.
- It ensures that natural laws remain symmetrical, allowing for conservation of energy, momentum, and charge.
- It serves as the mathematical and conceptual foundation for motion, forces, electromagnetism, and even thought processes.
- Despite the fragmentation of scientific theories, zero sustains mathematics as a universal

truth because it represents nature's deepest symmetry.
- As we continue to explore reality, we are not just discovering new knowledge—we are aligning ourselves with nature's hidden patterns, unknowingly evolving within the structured blueprint of existence.

In the next section, we will explore zero's deeper role in energy transformations—how it serves as the ultimate neutral state from which all physical and conceptual structures emerge, bridging existence and non-existence.

4.3 Zero in Energy Transformations – The State of Perfect Balance

Energy in the universe never appears or disappears—it only transforms. However, these transformations do not occur randomly; they follow precise laws of conservation, symmetry, and balance. At the core of these principles lies zero, the ultimate neutral reference point for all energy exchanges.

Zero is not merely an absence of energy; rather, it is the perfect equilibrium state that defines how energy is structured, exchanged, and redistributed. Every transformation in nature—from motion to charge interactions, from quantum mechanics to thermodynamics—operates around a zero-based equilibrium. This ensures that energy remains balanced rather than accumulating infinitely or vanishing unpredictably.

In this section, we explore how zero serves as the perfect neutral state in energy flow, maintains equilibrium between positive and negative energy interactions, and provides a universal reference point for measuring all physical transformations.

Zero as the Perfect Neutral State in Energy Flow

All energy interactions oscillate around a balanced state. Whether in classical physics, quantum mechanics, or thermodynamics, zero represents the ideal equilibrium condition where forces cancel out, and stability is maintained.

1. Zero as the Ground State in Quantum Mechanics

- In quantum mechanics, zero-point energy refers to the lowest possible energy state a system can have—even at absolute zero temperature.
- This proves that even "emptiness" is never truly empty—fluctuations in energy persist, ensuring that zero is never a void but a stable state of equilibrium.
- The Heisenberg Uncertainty Principle further reinforces that zero is not a static condition but a dynamic equilibrium.

2. Zero as the Rest State in Classical Mechanics

- In Newtonian physics, a system at zero net force is in perfect equilibrium.
- If no external force acts, an object remains at rest or moves at a constant velocity—meaning motion itself is measured relative to a zero-energy state.
- This principle governs everything from planetary orbits to atomic motion, ensuring energy remains structured rather than chaotic.

3. Zero in Thermodynamic Equilibrium

- In thermodynamics, the concept of absolute zero (0 K) represents the point where molecular motion ceases, marking the lowest possible energy state.
- Even though absolute zero cannot be physically reached, all temperature scales reference this theoretical zero-point equilibrium.
- This shows that heat and entropy always seek balance around a reference zero, preventing infinite energy buildup or complete dissipation.

From quantum to macroscopic levels, zero is the natural reference for all energy states, ensuring that transformations remain structured rather than disorderly.

The Balance Between Positive and Negative Energy Interactions

Energy does not exist in isolation—it always operates in dualities. Whether in electric charge, matter and antimatter, or wave functions, energy maintains balance by interacting symmetrically between positive and negative states.

1. Positive and Negative Charges in Electromagnetic Interactions

- In electromagnetism, zero represents electrical neutrality—every positive charge has a corresponding negative charge, ensuring no net charge buildup.
- Atoms remain stable because of the precise balance between positively charged protons and negatively charged electrons.
- Without this charge symmetry, matter itself would collapse into instability.

2. Matter and Antimatter Symmetry

- In particle physics, the existence of matter is balanced by its counterpart, antimatter.
- When matter and antimatter meet, they annihilate, returning their energy to the zero equilibrium state.
- This suggests that the universe itself is structured around a zero-net energy balance.

3. Energy Oscillations in Waves and Vibrations

- In sound waves, electromagnetic waves, and mechanical vibrations, energy always oscillates symmetrically between peaks (positive) and troughs (negative), with zero as the resting point.
- The fact that waves naturally return to a zero-resting position proves that energy does not build indefinitely—it always stabilizes through symmetry.

Across all scales, from atomic particles to wave dynamics, energy transformations follow structured symmetry, ensuring that zero remains the foundation of stability.

Zero as the Reference Point for Measuring All Physical Transformations

Without zero, we would have no way to define changes in energy, position, or force. Zero acts as the absolute reference point for measuring all physical transformations, ensuring consistency in scientific calculations and observations.

1. Zero in Motion and Acceleration

- In physics, all motion is relative to a reference frame, often defined at zero.
- Acceleration is measured by deviation from zero velocity.
- Gravitational potential is defined relative to a zero-energy baseline.

2. Zero in Thermodynamics and Energy Flow

- Entropy (disorder) always moves toward a stable equilibrium, which is mathematically referenced against zero net energy change.
- Heat transfer always equalizes toward a balanced state, preventing extreme temperature imbalances.

3. Zero in Time and Space Measurements

- Time is measured from a zero reference point, whether in relativity, cosmology, or classical mechanics.
- Space-time curvature is calculated relative to a zero gravitational field, ensuring stable cosmological models.

Thus, zero is not just a mathematical concept—it is the essential stabilizing reference point that defines all physical transformations.

In conclusion, Zero is the state of perfect balance in energy flow, ensuring that all physical processes, transformations, and interactions remain structured.

- It is the lowest energy state in quantum mechanics, preventing randomness and ensuring order.
- It is the neutralizing force in electromagnetic and matter-antimatter interactions, maintaining balance in atomic and cosmic structures.
- It serves as the absolute reference point in physics, thermodynamics, and wave dynamics, allowing us to define transformations in motion, temperature, and charge.

Ultimately, zero is not nothingness—it is the structured equilibrium upon which the entire universe is built.

This realization leads us to a deeper question:

If zero governs energy transformations at every level, could it also be the fundamental principle behind the existence of all reality?

In the next section, we will explore how zero is the key to understanding conservation laws, the stability of existence, and the fundamental symmetries that govern the universe.

4.4 Zero and the Conservation Laws of Physics

Zero is not merely a mathematical placeholder—it is the fundamental principle that ensures balance in all physical laws. The conservation laws of energy, momentum, and charge operate on the premise that everything in nature follows structured transformations, never creating or destroying energy, but merely redistributing it.

At the core of these conservation laws is zero—the equilibrium point that governs all physical interactions. Whether in classical mechanics, relativity, or quantum physics, zero ensures that every energy exchange, motion, and force follows precise symmetrical laws, maintaining the structured integrity of the universe.

In this section, we explore how zero is mathematically linked to conservation laws, plays a role in fundamental physics, and guarantees that all energy transitions remain balanced and symmetrical.

How the Conservation of Energy, Momentum, and Charge is Mathematically Linked to Zero

The three fundamental conservation laws in physics—energy, momentum, and charge—all depend on zero as their reference point. These laws dictate that the total quantity of energy, motion, and charge in a closed system remains constant, ensuring stability in physical interactions.

1. The Conservation of Energy and the Role of Zero

The First Law of Thermodynamics states:

Energy can neither be created nor destroyed, only transformed from one form to another.
However, if energy is always transforming, what prevents it from accumulating infinitely or dissipating into nothingness? The answer is zero.

- In any closed system, the total net energy remains constant, meaning that energy gain in one area must be counterbalanced by energy loss elsewhere.
- Even in nuclear reactions or high-energy collisions, the total energy before and after an event always sums to zero, ensuring perfect conservation.
- In the expansion of the universe, gravitational energy and kinetic energy counterbalance each other, maintaining a net zero-energy equilibrium.

Thus, zero is the mathematical anchor that ensures the conservation of energy remains valid in all physical transformations.

2. The Conservation of Momentum and Zero as the Reference Point

Newton's laws of motion and momentum conservation state that:
The total momentum of an isolated system remains constant unless acted upon by an external force.

- If two objects collide, their combined momentum before and after the collision always equals zero when measured relative to the system's center of mass.
- In rocket propulsion, the backward force of expelled gases is precisely counterbalanced by the forward motion of the rocket, ensuring momentum remains conserved.
- In astrophysics, planets orbit their stars while balancing gravitational pulls to maintain zero net momentum drift.

Momentum conservation only works because zero provides the absolute reference point for measuring changes in motion.

3. The Conservation of Charge and Zero as the Neutral State

Charge conservation states that:
The total charge in a closed system remains constant over time.

- Electrons and protons exist in equal numbers, ensuring that atoms are electrically neutral with a net charge of zero.
- Matter and antimatter interactions result in annihilation, returning energy to a zero-state equilibrium.
- In electrical circuits, charge movement follows a closed-loop principle, ensuring that no charge is lost or created, maintaining neutrality.

Without zero as the reference for neutrality, charge imbalances would result in instability, disrupting the very fabric of atomic and molecular structures.

The Role of Zero in Newtonian Mechanics, Relativity, and Quantum Physics

Zero is not just an abstract concept—it is a fundamental requirement for physical theories to remain valid across different domains of physics. Whether in classical mechanics, Einstein's relativity, or quantum mechanics, zero serves as the unifying principle that preserves conservation laws.

1. Zero in Newtonian Mechanics

Newton's laws of motion and gravity are structured around zero as the equilibrium condition.

- An object remains at rest (zero acceleration) unless acted upon by a force.
- For every action, there is an equal and opposite reaction (zero net force in a system).
- Equations of motion are derived from initial conditions where zero represents rest, balance, or symmetry.

Without zero, we would have no way to define rest, equilibrium, or the conservation of forces in classical physics.

2. Zero in Einstein's Theory of Relativity

Einstein's famous equation, $E = mc^2$, shows that mass and energy are interchangeable—but in a perfectly balanced manner.

- The curvature of space-time is measured relative to zero curvature in flat space, ensuring that general relativity remains mathematically consistent.
- The speed of light (c) remains constant, serving as the zero reference point for all relativistic transformations.
- Time dilation and length contraction occur in a way that preserves zero net energy loss in the system.

Thus, even in the vast cosmic framework, zero is the key mathematical structure that maintains relativity's predictive power.

3. Zero in Quantum Mechanics and Particle Physics

At the quantum level, zero remains essential in defining probabilities, energy states, and fundamental interactions.

- Zero-point energy is the lowest possible energy state a system can have, proving that even "nothingness" still contains structured fluctuations.
- The wave function of a quantum particle collapses to zero probability when observed,

ensuring that quantum mechanics remains mathematically rigorous.
- The vacuum state of the universe is defined by zero-field energy, meaning that even empty space contains fluctuations governed by zero-based symmetry.

Across every scale of physics, zero provides the foundation for defining rest states, equilibrium, and energy stability.

Why Zero Ensures That Energy Transitions Follow Precise Symmetrical Laws

Nature does not allow uncontrolled energy imbalances. Every process—from motion to thermodynamics, from charge interactions to quantum mechanics—follows a structured path that always returns to zero as the balancing reference.

This means that:

- Every force is counteracted by an equal and opposite force.
- Every energy gain is matched by an equivalent loss elsewhere.
- Every charge interaction maintains perfect neutrality over time.

Mathematically, zero ensures that equations governing energy transformations remain symmetrical.

- In mechanics, forces balance around zero, ensuring stability.
- In thermodynamics, entropy follows a symmetrical path toward zero equilibrium.
- In electromagnetism, wave functions oscillate symmetrically around a zero-rest state.

Ultimately, zero is the hidden symmetry that structures all conservation laws, preventing disorder while allowing transformations to occur within defined constraints.

In conclusion, Zero is not an arbitrary number—it is the foundation of conservation laws that govern all of physics.

- It ensures that energy, momentum, and charge remain balanced, preventing infinite accumulation or loss.
- It serves as the reference point in Newtonian mechanics, relativity, and quantum physics, maintaining mathematical consistency across different scales.
- It guarantees that all energy transitions follow precise symmetrical laws, reinforcing the stability of the universe.

Without zero, the conservation laws would collapse, and the structured nature of physics would break down. However, because zero remains constant, the universe functions as a self-regulating system where energy transformations always return to equilibrium.

This brings us to a deeper realization:
If zero is the stabilizing force of energy conservation, could it also be the guiding principle behind existence itself?
In the next section, we will explore how zero forms the foundation of existence and non-existence—explaining why reality itself is structured around the balance between something and nothing.

4.5 Zero as the Foundation of Existence and Non-Existence

Zero is not just a number—it is a paradox, a bridge between nothingness and existence. It represents both the void and the source of everything, forming the fundamental boundary between what exists and what does not. Throughout history, zero has been viewed philosophically as emptiness, scientifically as equilibrium, and mathematically as the foundation of structure. But at its deepest level, zero is more than a concept—it is the ultimate reference point that defines the very nature of reality.
As we have seen, zero governs conservation laws, symmetry, and energy balance. But beyond these, it also defines time, entropy, and the structure of the universe itself. In this section, we explore how zero represents the paradox of existence, serves as the dividing line between being and non-being, and explains the deep connection between entropy, time, and cosmic evolution.

The Philosophical Paradox of Zero as Both Nothingness and the Origin of Everything

Zero has always puzzled both scientists and philosophers because it embodies a contradiction:

- Mathematically, zero represents nothingness, an empty state, a void.
- Physically, zero represents perfect equilibrium, the balanced state of all energy and forces.
- Philosophically, zero represents both the absence of existence and the origin from which everything emerges.

How can zero be both the void and the source? The answer lies in energy equilibrium and symmetry.

- The Big Bang is often described as an event that emerged from "nothing," but in reality, it emerged from a perfectly balanced state (zero net energy).
- In quantum physics, virtual particles continuously appear and disappear, fluctuating around a zero-energy vacuum state.
- Even in information theory, zero represents the absence of data, yet it is also required to define all binary logic (0s and 1s).

Thus, zero is not a simple absence—it is the neutral foundation upon which existence is structured. It allows energy fluctuations, defines mathematical frameworks, and sustains the laws of physics. Without zero, the transition between nothing and something would have no reference point.

Why Zero is the Reference Point Between Existence and Non-Existence

Zero is the mathematical and physical boundary between being and non-being. Everything that exists is measured relative to zero as the fundamental baseline.

- Energy exists relative to a zero-point vacuum state.
- Motion is measured relative to zero velocity (rest).
- Time progresses relative to a starting zero-moment (initial conditions).

If zero did not exist as a reference, there would be no way to define existence itself. The act of measuring, perceiving, and even understanding anything requires a point of comparison, a balance between presence and absence.

1. Zero in the Structure of the Universe

- The universe's total energy may be exactly zero—the sum of all mass-energy and gravitational potential cancels out.
- Quantum fields fluctuate around a zero vacuum state, producing all known particles and interactions.
- At the largest scale, the universe expands around a balanced zero-net energy state, ensuring long-term stability.

2. Zero in Life and Consciousness

- Life itself depends on metabolic balance, where energy input and output must stabilize toward zero net energy loss.
- Human consciousness operates through oscillations between neural excitations and inhibitory responses, all structured around balanced zero-net neural activity.
- Cognitive processes rely on contrast (light vs. dark, sound vs. silence, presence vs. absence), proving that zero is essential for perception itself.

Thus, zero is not an abstract concept—it is the very reference point that allows all definitions of existence.

How Zero Connects with Entropy, the Universe's Expansion, and the Nature of Time

Zero plays a critical role in time, entropy, and the evolution of the universe. One of the most profound realizations in physics is the inverse relationship between entropy and time.

1. Time Emerges from Zero

- Time is not an absolute quantity—it is an emergent property that arises from the presence of order and low entropy.
- In the early universe, entropy was extremely low, allowing time to move forward in a structured way.
- At maximum entropy (complete disorder), time numerically becomes zero, meaning it ceases to exist.

This suggests that time is fundamentally linked to the order of the universe, with zero serving as the ultimate reference point.

2. The Inverse Relationship Between Entropy and Time

Entropy is the measure of disorder in a system. The Second Law of Thermodynamics states that entropy always increases over time. However, this also means:

- When entropy is low, time flows in a structured manner.
- As entropy increases, time becomes less distinguishable, leading toward a state where time effectively stops.
- At infinite entropy (complete disorder), time is numerically zero—it does not exist.

This is why in black holes (where entropy is maximized), time slows down and eventually ceases to have meaning. It is also why, at the heat death of the universe (where all energy is maximally distributed), time would effectively reach a zero state.

Thus, zero is not just a reference for energy, charge, and existence—it is also the reference for time itself.

3. The Universe Expands Around Zero as the Reference State

- The expansion of the universe is driven by an equilibrium between energy and gravitational pull.
- As the universe expands, it moves toward a high-entropy state, approaching an equilibrium where time, structure, and motion cease to exist.
- At the end of the universe's lifespan, zero will once again define the boundary between existence and non-existence.

Zero is the ultimate reference point, the bridge between existence and non-existence.

- It represents the paradox of being both nothingness and the source of everything.
- It defines the boundary between existence and non-existence, ensuring that all physical and conceptual systems have structure.
- It governs entropy, the expansion of the universe, and the nature of time itself.

Ultimately, zero is the key to understanding why reality exists at all. It is the state of perfect balance, the hidden symmetry behind all transformations, and the guiding principle that structures energy, space, and time.

This leads us to a deeper realization:

If zero is the foundation of time, entropy, and the universe's structure, could it also be the underlying force behind intelligence, evolution, and the emergence of complexity?

In the next section, we will explore how zero integrates into Universal Energy Dynamics (UED), revealing why it is the mathematical, physical, and philosophical key to understanding the deepest mysteries of existence.

4.6 Zero in Universal Energy Dynamics (UED)

Zero is not merely a numerical abstraction—it is the ultimate balancing mechanism that governs all physical, biological, and cognitive processes. In the framework of Universal Energy Dynamics (UED), zero is the reference state toward which all energy transformations, forces, and interactions stabilize. It represents the fundamental equilibrium that ensures order, prevents chaotic instability, and allows structured evolution in the universe.

Through our exploration of zero in physics, conservation laws, entropy, and time, it has become evident that all energy fluctuations must return to a state of balance. Whether in the dynamics of galaxies, subatomic particles, or human cognition, zero emerges as the invisible force that sustains equilibrium across all domains.

In this section, we explore how zero integrates into UED as the stabilizing principle of reality, why all energy fluctuations must return to zero-equilibrium, and how this foundational principle extends into cosmic structure, intelligence, and the deeper nature of existence.

How Zero Integrates into UED as the Ultimate Balancing Mechanism

Universal Energy Dynamics (UED) is based on the premise that energy is the fundamental reality of existence. Everything—matter, force, thought, and even consciousness—is a transformation of energy within a structured framework. However, for this system to remain self-sustaining, energy interactions must be governed by a balancing force. That force is zero.

1. Zero as the Foundation of Energy Equilibrium

All energy exchanges follow structured patterns—they never continue indefinitely in one direction but eventually return to a zero-balanced state.

- In physics, energy states oscillate around equilibrium, ensuring that no system gains or loses energy infinitely.
- In biological systems, metabolic processes stabilize toward homeostasis, keeping organisms in a zero-net energy state over time.
- In cognitive processes, neural activity oscillates between excitation and inhibition, maintaining mental stability through balance.

Thus, zero is the invisible regulator that ensures energy remains structured rather than chaotic. Without it, the universe would either collapse into disorder or stagnate without transformation.

2. The Role of Zero in Energy Transformation and Flow

- In thermodynamics, heat always moves from hot to cold, seeking zero temperature difference (thermal equilibrium).
- In wave mechanics, oscillations always return to a resting zero position, ensuring stability in sound, light, and motion.
- In gravitational interactions, celestial bodies stabilize into orbits where forces cancel out to near-zero net acceleration.

This confirms that zero is not a passive absence of energy—it is the dynamic reference point that guides the structured flow of all energy interactions.

Why All Energy Fluctuations Must Return to Equilibrium (Zero Reference)

One of the core principles of UED is that energy cannot remain in a state of imbalance indefinitely. All systems—whether physical, biological, or cognitive—naturally return to a zero-state equilibrium.

1. Zero in Classical Physics: Motion and Force Equilibrium

- Newton's First Law of Motion states that an object will remain in uniform motion or at rest unless acted upon by an external force.
- However, in real-world conditions, friction, resistance, and external interactions always guide objects back to a state of rest (zero net force).
- Even in planetary motion, gravitational and centrifugal forces balance each other, ensuring a near-zero net acceleration over long timeframes.

2. Zero in Quantum Mechanics: Wave Collapse and Probability Distributions

- In quantum physics, wave functions describe probabilities, oscillating between possible states.

- When measured, the wave function collapses to a definite value, effectively returning to a zero-probability superposition.
- This suggests that even in probabilistic systems, energy transitions seek a zero-based reference.

3. Zero in Biological and Neural Systems: Homeostasis and Stability

- The human body maintains stable internal conditions through homeostasis, ensuring a zero-net metabolic energy balance over time.
- The nervous system operates through excitatory and inhibitory signals, ensuring neural equilibrium and mental stability.
- Even consciousness itself functions through a balance of sensory inputs and cognitive processing, preventing chaotic overload.

Thus, zero acts as the stabilizing principle that ensures all dynamic systems return to equilibrium, preventing runaway instability.

How Zero Serves as the Foundation for Understanding Cosmic Structure and Intelligence

If zero governs physical forces, energy transformations, and biological systems, could it also be the fundamental principle behind cosmic evolution and intelligence? UED suggests that zero is not only a balancing mechanism but also the foundational structure upon which intelligence and cosmic order emerge.

1. Zero in Cosmic Evolution and the Structure of the Universe

- The universe's large-scale structure follows gravitational equilibrium patterns, ensuring that galaxies form stable clusters rather than collapsing chaotically.
- Dark matter and dark energy balance the expansion of the universe, maintaining a near-zero net energy framework.
- Cosmic background radiation remains at an equilibrium temperature close to absolute zero (2.7 K), preserving stability in universal evolution.

This suggests that zero is the guiding principle that allows the universe to evolve while remaining structured.

2. Zero as the Root of Intelligence and Thought

- Intelligence operates by recognizing patterns, balancing contradictions, and processing

structured energy.
- Neural networks in the brain function through zero-based equilibrium, with electrical activity oscillating between excitation and inhibition.
- Even in artificial intelligence, decision-making relies on weighing positives and negatives, converging toward balanced conclusions (zero net uncertainty).

Thus, zero is not just a physical principle—it is also the foundation of structured cognition, allowing intelligence to emerge in both biological and artificial systems.
In conclusion, Zero is the ultimate balancing force in Universal Energy Dynamics (UED).

- It ensures that all energy transformations stabilize over time, preventing disorder.
- It serves as the reference state for motion, thermodynamics, quantum fluctuations, and biological homeostasis.
- It provides the foundation for understanding cosmic structure and intelligence, proving that balance is the fundamental rule of existence.

Without zero, energy would either dissipate into chaos or stagnate without transformation. But because zero exists as the guiding force of equilibrium, the universe, life, and intelligence can evolve while maintaining structure.
This brings us to a deeper question:
If zero is the fundamental stabilizing force, could it also be the key to unlocking the deeper purpose of existence, intelligence, and the structured evolution of the cosmos?
In the next section, we will expand our understanding of zero, exploring its role in defining infinity, transformation, and the interconnected fabric of the universe.

4.7 Expanding Beyond Zero – The Infinite Possibilities

Zero is not the end of a concept—it is the gateway to infinity. It is the state of perfect balance from which limitless transformations emerge. Throughout our exploration, we have seen how zero governs energy equilibrium, stabilizes physical laws, and structures the evolution of the universe. But beyond serving as a reference for balance, zero also defines the foundation for infinite possibilities.
In Universal Energy Dynamics (UED), zero is not an absolute void but a dynamic state of equilibrium that allows the structured flow of energy, consciousness, and existence itself. The universe does not remain in a static zero state; rather, it cycles through zero-equilibrium into structured transformations and back, allowing continuous creation, evolution, and intelligence to emerge.
This section explores how zero serves as the mathematical and physical gateway to infinity, how the universe oscillates between zero-balanced states and structured reality, and how we are preparing to extend this concept into intelligence, evolution, and the deeper nature of existence.

How Zero is the Gateway to Infinity, Defining Limitless Transformations

In many ways, zero and infinity are interconnected. Zero represents the state of perfect balance, while infinity represents the boundless transformations that emerge from it.

1. Zero as the Seed of Infinite Expansion

- In calculus, zero is the foundation for infinitesimals—infinitely small quantities that allow the definition of limits, derivatives, and integrals.
- In set theory and number systems, zero is required to define negative numbers, fractions, and the entire real number line, extending toward infinity.
- In physics, zero represents the lowest energy state, but quantum fluctuations ensure that even from this state, transformations occur infinitely.

Thus, zero is not an endpoint—it is the neutral foundation from which infinite transformations arise.

2. Zero and the Expansion of the Universe

- The universe originated from a zero-net energy state, yet it continuously expands, allowing infinite transformations of matter, energy, and space.
- Even though space expands, it does not accumulate infinite energy; instead, it redistributes energy symmetrically, preserving the zero-energy framework of the cosmos.
- The interplay between dark matter and dark energy ensures that the universe follows structured cycles of expansion and equilibrium, rather than collapsing into disorder.

Thus, zero does not represent stillness; rather, it is the dynamic equilibrium from which infinite cosmic cycles emerge.

Why the Universe Cycles Through States of Zero-Equilibrium to Structured Existence

One of the greatest mysteries in physics is why the universe does not remain in a perfectly balanced state forever. If zero represents absolute equilibrium, then why do structures, galaxies, intelligence, and life emerge?

The answer lies in the continuous oscillation between zero and structured existence.

1. The Universe as a Balanced Oscillating System

- In wave physics, oscillations always return to a zero-resting position before creating the next cycle.
- In quantum mechanics, particles continuously fluctuate in and out of zero-energy states, creating dynamic interactions.

- In cosmology, the universe undergoes cycles of equilibrium and transformation, ensuring that it neither collapses nor remains static.

This suggests that the universe is not a one-time event but a continuous process of structured transformations arising from zero-equilibrium.

2. Entropy, Time, and the Zero-Reset Mechanism

- We previously discussed how time is linked to entropy, with time numerically approaching zero at infinite entropy.
- This implies that after the universe reaches a maximum state of entropy, a new cycle begins from zero, allowing structured existence to restart.
- In physics, models like the cyclic universe, conformal cosmology, and quantum loop gravity suggest that the universe may repeat cycles of zero-equilibrium to structured existence.

Thus, zero is not a static void—it is a dynamic, self-regulating mechanism that allows infinite structured cycles of transformation.

Preparing to Explore How Zero Extends into Intelligence, Evolution, and Reality Itself

If zero serves as the foundation for structured cosmic cycles, could it also be the fundamental principle behind intelligence, consciousness, and evolution?

1. Zero and the Evolution of Intelligence

- Intelligence emerges as a process of balancing and optimizing information, decisions, and energy.
- The brain operates through zero-net equilibrium of excitatory and inhibitory signals, allowing structured thought and decision-making.
- Even artificial intelligence follows optimization algorithms that converge toward zero error, proving that learning itself follows a zero-based equilibrium principle.

Thus, zero is not just a mathematical or physical law—it is also the key to understanding intelligence and thought processes.

2. Zero and the Hidden Structure of Reality

- If the universe follows zero-equilibrium cycles, then all of existence operates under a hidden order of balance and transformation.
- This suggests that even the emergence of life, consciousness, and decision-making are not

random but structured by zero as the fundamental reference point.
- Philosophically, zero represents both the stillness of the void and the foundation of infinite creation, bridging science and metaphysical concepts.

This realization leads us to the next phase of our exploration—how zero integrates into the very fabric of intelligence, evolution, and human understanding.
In conclusion, Zero is not an endpoint—it is the foundation of limitless possibilities.

- It defines the boundary between balance and transformation, allowing infinite expansion.
- The universe does not remain static at zero—it cycles through equilibrium and structured existence.
- This same principle extends into intelligence, evolution, and the deeper nature of reality.

Zero is the hidden structure behind energy, time, and intelligence. It allows the universe to self-regulate, evolve, and continuously generate complexity.
This brings us to our next profound question:
If zero is the key to energy balance, cosmic cycles, and intelligence, could it also be the blueprint for consciousness, decision-making, and the nature of self-awareness?
In the next chapter, we will expand our understanding of zero beyond physics, exploring its role in intelligence, thought processes, and the structure of consciousness itself.

Answers to Big Questions

(FAQ for Part 1)

To further strengthen our framework, this section will address some of the most fundamental questions that naturally arise from Part 1 of The Blueprint. These questions will help readers connect with the core ideas of UED, resolve potential doubts, and prepare them for deeper explorations in upcoming chapters.

1. Why do we need a Universal Blueprint?

Science has advanced tremendously in understanding the physical universe, life, and intelligence, yet it remains fragmented. Physics explains forces, biology deciphers life, and philosophy contemplates existence—but they remain disconnected.

A Universal Blueprint is necessary to:

- Unify all realms of knowledge—from physics to consciousness—under a single framework.
- Reveal the hidden order in nature, ensuring that energy, symmetry, and intelligence are understood as part of a larger system.
- Bridge science and philosophy, allowing us to answer deeper existential questions with logical consistency.

2. Why is energy considered the fundamental reality in UED?

In Universal Energy Dynamics (UED), energy is the core substance from which everything emerges.

- Matter is a condensed form of energy (Einstein's E=mc2E=mc^2E=mc2).
- Forces are energy interactions—gravity, electromagnetism, and nuclear forces are all different ways energy manifests.
- Even thought and consciousness emerge from energy—the structured flow of electrical and biochemical signals in the brain.

Thus, energy is not just one aspect of reality—it is reality itself. Everything we perceive is a transformation of energy within structured systems.

3. How does symmetry shape the universe?

Symmetry is the hidden order behind everything in nature. It ensures that:

- Forces and interactions remain balanced—Newton's third law (action-reaction) follows symmetrical principles.
- Matter and antimatter exist in pairs—ensuring conservation of energy and charge.
- Biological evolution prefers symmetry—because it provides efficiency and stability.
- Thought and decision-making operate through symmetry—balancing emotions, logic, and memory in structured ways.

Without symmetry, the universe would be chaotic, imbalanced, and unpredictable. It is the mechanism that ensures stability while allowing transformation.

4. Why is zero the most important concept in physics and philosophy?

Zero is the ultimate reference point for reality because:

- It defines balance and neutrality—all forces and energies stabilize around zero.
- It governs conservation laws—momentum, energy, and charge always sum to zero in closed systems.
- It represents the foundation of time and entropy—time emerges from zero entropy and ceases when entropy is maximized.
- It is both nothing and everything—zero is paradoxically the source of existence and the state to which everything returns.

Thus, zero is not an empty void—it is the foundation of equilibrium that governs the entire universe.

5. Does the universe have a net zero energy balance?

Yes, many physicists propose that the universe has a total energy of zero:

- Matter contains positive energy, while gravitational fields contain negative energy.
- The sum of all mass-energy and gravitational potential could be exactly zero.
- This means that the universe could have emerged from a quantum fluctuation in a zero-energy vacuum, requiring no external force.

Thus, the universe is not created from "nothing" but from a state of perfect balance, where zero-energy fluctuations allow structured reality to emerge.

6. How is zero linked to time, entropy, and the future of the universe?

Time and entropy share an inverse relationship, meaning:

- Time moves forward as entropy increases—this is why we experience the "arrow of time."
- At maximum entropy (complete disorder), time numerically reaches zero, meaning it ceases

to exist.
- The universe may reach a state where all energy is evenly distributed, halting motion, structure, and time itself.

This suggests that the universe cycles between structured existence and zero-equilibrium states, leading to infinite cycles of transformation.

7. Does UED challenge existing scientific theories?

No. UED does not contradict physics, biology, or neuroscience—it expands on them by unifying their principles.

- It aligns with conservation laws, relativity, and quantum mechanics, explaining them in a broader framework.
- It integrates biological and cognitive sciences, showing that life and intelligence follow the same principles of energy flow.
- It connects philosophy with physics, providing a structured explanation for existence, time, and consciousness.

Thus, UED is not a replacement for existing science but a synthesis of knowledge, offering a unified perspective.

8. What does this mean for human intelligence and consciousness?

If energy and symmetry define everything in the universe, then:

- Consciousness is an emergent property of structured energy flow.
- Thoughts follow the same principles as wave interactions and quantum fluctuations.
- Decision-making operates through equilibrium—balancing inputs, experiences, and learning.

This suggests that intelligence itself is a structured transformation of energy, allowing awareness, memory, and reasoning to emerge in a zero-based equilibrium system.

9. If everything moves toward zero-equilibrium, does free will exist?

Yes, but it operates within a structured system.

- Free will is not absolute randomness, but the ability to navigate structured energy pathways.
- The brain operates on predictable neural energy exchanges, but consciousness can alter energy flow through intentional decisions.
- This means we are not separate from universal energy dynamics—we are an extension of it.

Thus, free will exists, but it is governed by the same laws of symmetry, balance, and transformation that define the universe.

10. Where does this take us next?

Having established energy, symmetry, and zero-equilibrium as the foundation of reality, we now prepare to explore intelligence, evolution, and the deeper nature of existence.

- How does intelligence emerge from energy transformations?
- Can consciousness be fully explained within this framework?
- What role does zero play in self-awareness and the perception of reality?

These questions will guide us into the next phase of our exploration, where UED extends beyond physics into the very essence of thought, perception, and existence.
In conclusion, The Universal Blueprint is not just about physics—it is about understanding the entire structure of reality, from fundamental forces to intelligence itself.

- Energy is the core of existence.
- Symmetry ensures balance and order.
- Zero defines the boundary between existence and non-existence.

With this foundation in place, we are now ready to explore the role of intelligence, self-awareness, and the next stages of universal evolution.
This is where our journey continues—into the structure of thought, consciousness, and the deep nature of reality itself.

Are We in a Simulation?

A Perspective from Universal Energy Dynamics (UED)

The simulation hypothesis suggests that our reality is a computational construct, a simulated environment created by an advanced intelligence. This idea has gained attention due to rapid advancements in artificial intelligence, computing, and physics. However, does this align with the principles of Universal Energy Dynamics (UED)? Or is the simulation theory just a philosophical illusion?

Let's analyze the simulation hypothesis through the lens of energy, symmetry, zero-equilibrium, and intelligence.

1. What Does Simulation Actually Mean?

In digital terms, a simulation is an artificial environment programmed within a computational system. If our universe were a simulation:

- It would require a computing substrate (a higher-dimensional or external system running it).
- It would operate based on programmed rules, similar to a video game or virtual reality.
- Intelligence, consciousness, and physics would be emergent properties within the simulation, much like non-player characters (NPCs) in a game.

But if the entire universe is a simulation, then what exists outside of it? And more importantly, what is the fundamental substance of this so-called simulation?

2. UED Perspective: The Universe is Structured, But Not a Simulation

Universal Energy Dynamics (UED) argues that:

- Energy is the fundamental reality. Everything—matter, thought, forces—is a transformation of energy.
- Symmetry governs all interactions. The universe does not behave randomly but follows structured transformations.
- Zero-equilibrium regulates all processes. All systems return to balance rather than being artificially maintained by an external force.

If the universe were a simulation, then what is being simulated? And more importantly—who or what is running it?

Unlike a programmed simulation, UED suggests that the universe is self-regulating through energy, balance, and structured transformations. There is no need for an external "programmer" or "computer" because the universe operates through fundamental laws that sustain and evolve naturally.

3. The Laws of Physics Are Not Arbitrary Like a Simulation

In a simulation, laws can be changed by the programmer—they are artificial constraints.

- If we were in a simulation, we should be able to find instances where physics "breaks" or where inconsistencies appear.
- However, across all observations, the universe follows strict energy conservation, symmetry, and mathematical consistency.

This suggests that our reality is not an arbitrary computational construct but a self-sustaining system based on fundamental principles.

4. Can Consciousness Exist in a Simulation?

If we assume that consciousness emerges from energy transformations in the brain, then in a simulated universe:

- Consciousness would be a programmed illusion rather than a fundamental experience.
- Decisions would be predefined by algorithms rather than arising from structured energy interactions.
- Free will would be a computational process with no real autonomy.

However, consciousness in our universe behaves unpredictably—it is not a scripted event.

- Neural processes follow energy flow and symmetry, not predetermined instructions.
- Humans can create, imagine, and break patterns, which is beyond typical computation.
- Quantum mechanics introduces fundamental uncertainty, which contradicts strict determinism in a simulated environment.

Thus, if intelligence and consciousness are not artificial, it suggests that we are not in a simulated construct but part of an organic, structured energy system.

5. If Not a Simulation, Then What?

Rather than viewing reality as a simulation, UED proposes that existence is an evolving energy system governed by structured transformations.

- Energy is not "simulated"—it is the real substance of everything.
- Symmetry and balance are not programmed rules—they are intrinsic to the way energy self-regulates.
- Zero is not a computational trick—it is the natural reference point for all transformations.

If we were in a simulation, there should be evidence of artificial boundaries, external intervention, or "code-breaking" anomalies—but the universe behaves as a self-contained, continuous system.
Thus, instead of being a simulation, the universe is an emergent structure that follows energy equilibrium and transformation dynamics.

Conclusion: We Are Not in a Simulation—We Are in an Evolving Energy System

The universe does not require an external computer, programmer, or artificial code.

- It operates through self-regulating energy transformations, not predefined simulation rules.
- Consciousness follows structured energy flow, not programmed algorithms.
- There is no evidence of artificial constraints—only structured energy equilibrium.

Thus, rather than being a simulation, our existence is part of a deeper, self-sustaining universal framework that operates through energy, symmetry, and balance.
This leads to the next question:
If we are not in a simulation, what is the true nature of reality, and how does intelligence fit into the universal structure?
This is where we continue our exploration—into the deeper nature of intelligence, consciousness, and structured existence within UED.

PART 2: The Science behind the Blueprint

Chapter 5: UED and the Laws of Physics

In Part 1, we established Universal Energy Dynamics (UED) as the foundational principle that governs all existence. We explored how energy, symmetry, and zero-equilibrium structure reality, allowing for balance and transformation across physics, biology, and cognition. We also saw how zero serves as the ultimate reference point, ensuring that all energy fluctuations return to equilibrium.

Now, in Part 2, we take the next step—integrating UED into existing scientific laws. While traditional physics has provided us with Newtonian mechanics, relativity, quantum mechanics, and thermodynamics, these frameworks have remained fragmented, each explaining reality within its own domain. However, when viewed through the lens of UED, these laws are not separate but interconnected principles of structured energy flow.

This chapter explores how UED naturally aligns with classical and modern physics, showing that:

- Newton's Laws describe energy as motion and force balance.
- Relativity explains gravity as an energy distribution system rather than just space-time curvature.
- Quantum mechanics reflects energy balance through probability waves and entanglement.
- Thermodynamics redefines entropy as energy redistribution rather than mere disorder.

Rather than replacing existing physics, UED provides a unifying framework—one that reveals the underlying energy-based structure connecting all physical laws. By understanding energy as the fundamental reality behind motion, forces, and transformations, we will see how physics, like everything else, follows the universal blueprint of energy balance and symmetry.

5.1 How UED Naturally Fits into Classical and Modern Physics

Universal Energy Dynamics (UED) is not a new theory that replaces existing scientific laws—rather, it serves as a unifying principle that integrates the fragmented domains of physics into a single framework. While Newtonian mechanics, relativity, quantum mechanics, and thermodynamics each describe different aspects of reality, they all ultimately revolve around energy transformations, balance, and symmetry.

Traditional physics has often treated forces, motion, and energy as separate phenomena. However, when viewed through the lens of UED, we see that all physical laws are simply structured energy interactions governed by symmetry and zero-equilibrium. This means that the entire universe functions as an interconnected energy system, with all transformations occurring in structured, self-regulating cycles.

This section explores why UED is not a competing theory but a framework that naturally aligns with established physics, providing deeper insight into the nature of energy, symmetry, and equilibrium.

Why UED is Not a New Theory, But a Unifying Principle

Physics has historically been divided into separate domains:

- **Newtonian Mechanics** describes motion and forces in macroscopic systems.
- **Relativity** explains gravity and space-time curvature in high-energy and cosmic scales.
- **Quantum Mechanics** governs the behavior of subatomic particles.
- **Thermodynamics** deals with heat, entropy, and energy transformations.

Each of these frameworks has successfully explained different aspects of reality, but they remain disconnected. Newtonian mechanics breaks down at relativistic speeds, relativity struggles to integrate with quantum mechanics, and thermodynamics often treats entropy as disorder rather than energy balance.

UED bridges these gaps by identifying energy as the fundamental unifying reality and revealing how all physical laws are simply different manifestations of structured energy flow.

Key Unifying Aspects of UED:

- **Energy as the Core Reality** – Everything in physics can be understood as energy interacting through structured transformations.
- **Symmetry as the Hidden Order** – All physical laws obey symmetry principles, ensuring that energy remains structured, predictable, and balanced.
- **Zero-Equilibrium as the Foundation** – Every energy interaction follows a path toward equilibrium, ensuring conservation and stability.

This means that UED does not reject classical or modern physics—it simply connects them under a single framework, showing that energy, motion, and transformation are part of a continuous, structured system.

How Energy as the Core Reality Aligns with Newtonian Mechanics, Relativity, Quantum Mechanics, and Thermodynamics

Physics has long treated matter, forces, and energy as separate entities, but UED reveals that they are all manifestations of the same underlying energy structure.

1. Newtonian Mechanics: Energy as Motion and Force Balance

- **Newton's First Law** (inertia) states that an object in motion stays in motion unless acted upon by an external force.
- **Newton's Second Law** (F=ma) shows that force is a measure of how energy interacts with mass to produce acceleration.
- **Newton's Third Law** (action-reaction) follows **energy conservation principles**, ensuring that forces always balance symmetrically.

❖ **UED Perspective:** Newtonian mechanics is simply a structured description of energy transfer through force interactions. All motion is governed by energy conservation and equilibrium principles.

2. Relativity: Gravity as Energy Distribution, Not Just Space-Time Curvature

- Einstein's General Relativity describes gravity as the curvature of space-time due to mass-energy.
- However, UED interprets gravity as an energy distribution mechanism that follows equilibrium principles.
- Gravitational waves are simply energy redistributions that balance space-time geometry over large scales.

❖ **UED Perspective:** Rather than treating gravity as only a geometric distortion, UED sees it as an emergent energy balancing mechanism that aligns mass-energy distributions across the universe.

3. Quantum Mechanics: Probability Waves as Energy Balance

- Quantum mechanics describes particles as probability waves, with their exact positions determined only when measured.
- Wave-particle duality suggests that matter and energy exist in structured transformations.
- Quantum entanglement suggests that energy interactions are interconnected across space-time.

❖ **UED Perspective:** Quantum mechanics is not random—it is simply energy operating within structured balance rules, ensuring symmetry at subatomic scales.

4. Thermodynamics: Entropy as Energy Flow, Not Disorder

- Traditional thermodynamics often describes entropy as disorder increasing over time.
- However, UED redefines entropy as structured energy redistribution, ensuring that systems remain balanced.
- The heat death of the universe is not disorder but a state of zero-equilibrium, where all energy transitions stabilize.

❖ **UED Perspective:** Thermodynamics is simply the structured flow of energy toward balance, rather than an increase in disorder.

Why Symmetry and Zero-Equilibrium Provide a Common Framework for Explaining All Physical Laws

At every scale of physics, symmetry and zero-equilibrium govern all transformations.

- Newtonian mechanics follows zero-sum force interactions.
- Relativity maintains energy balance through gravitational fields.
- Quantum mechanics operates through probability distributions that maintain net zero-energy shifts.
- Thermodynamics ensures that energy redistributes toward equilibrium, preventing chaotic instability.

This suggests that all known physics follows structured energy balance principles, with zero as the stabilizing reference.

1. Symmetry as the Fundamental Rule of Physics

- Symmetry ensures that energy, motion, and transformations remain structured.
- Newton's action-reaction law follows symmetry principles.
- Quantum field theory describes forces as symmetrical energy exchanges.

❖ **UED Perspective:** Symmetry is the universal order that structures energy into predictable, evolving patterns.

2. Zero-Equilibrium as the Ultimate Stabilizer

- All energy interactions eventually return to a state of zero net force or energy balance.
- Gravity, electromagnetism, and nuclear forces follow equilibrium rules that prevent uncontrolled chaos.
- Quantum wave functions collapse to a zero-reference state, ensuring probability consistency.

❖ **UED Perspective:** Zero is not an absence—it is the ultimate energy equilibrium state that all systems follow.

UED as the Unifying Framework for All Physical Laws

Universal Energy Dynamics (UED) does not replace physics—it reveals the hidden connection between classical and modern physics by:

- Unifying Newtonian mechanics, relativity, quantum mechanics, and thermodynamics under a single energy-based framework.
- Demonstrating that energy, symmetry, and zero-equilibrium govern all transformations, preventing disorder while allowing structured evolution.
- Providing a deeper understanding of conservation laws, motion, gravity, and entropy, showing that all physical interactions are part of a continuous energy system.

Thus, rather than viewing physics as disconnected theories, UED provides a universal model where all forces, transformations, and interactions are structured through energy balance principles.
This leads us to the next question:
If UED naturally integrates into physics, can it also explain the emergence of intelligence, biological evolution, and cognitive processes as structured energy dynamics?
In the next section, we explore how Newton's Laws align with UED, showing that motion, force, and action-reaction principles are simply structured manifestations of energy equilibrium.

5.2 Newton's Laws: Energy as Motion and Force Balance

Newton's Laws of Motion are among the most fundamental principles in physics, describing how objects move and interact under forces. However, when viewed through the lens of Universal Energy Dynamics (UED), Newton's Laws reveal a deeper connection to energy equilibrium, symmetry, and structured transformation.
Rather than being separate mechanical rules, Newton's Laws are simply manifestations of energy conservation and structured force interactions. They describe how energy distributes, stabilizes, and transforms, ensuring that motion follows predictable patterns while maintaining balance in the universe.
In this section, we explore how Newton's First, Second, and Third Laws align with UED, demonstrating that motion, force, and momentum conservation are all structured energy transformations governed by equilibrium principles.

First Law: Inertia and Zero-Equilibrium – How Energy Balance Defines Motion and Rest

Newton's First Law of Motion states:

An object at rest remains at rest, and an object in motion continues in uniform motion unless acted upon by an external force.

This law is often described as inertia, but from the UED perspective, it is actually an expression of energy equilibrium.

1. Zero-Equilibrium in Motion and Rest

- A system at rest has zero net energy transformation—all forces acting on it are in equilibrium.
- Motion is not something separate from energy—it is a state where energy remains structured as kinetic movement rather than potential stability.
- An object in motion continues moving because no additional energy input is needed to sustain velocity—it already possesses kinetic energy.

Thus, motion and rest are not two different conditions, but two different equilibrium states of energy.

2. Inertia as an Expression of Energy Conservation

- Objects with greater mass have greater inertia, meaning they require more energy to change their motion.
- This aligns with UED's principle that mass is a structured form of energy, and more massive objects contain greater energy density.
- The idea that objects do not change motion unless influenced by an external force is simply a natural consequence of energy conservation and balance.

❖ **UED Perspective:** Newton's First Law is a direct manifestation of energy equilibrium—objects remain in their current state (rest or motion) unless an external force disrupts their balanced energy state.

Second Law: Force and Acceleration – Why Forces are Structured Energy Transformations

Newton's Second Law of Motion states:

The force acting on an object is equal to its mass times acceleration.

$F = ma$

This law defines how force interacts with mass to produce acceleration. However, from a UED perspective, this is simply a structured transformation of energy.

1. Force as an Energy Redistribution Mechanism

- When a force is applied, it transfers energy into an object, causing a change in motion.
- The equation F=ma shows that acceleration is directly proportional to the energy input per unit mass.
- Larger forces introduce greater energy transformations, resulting in higher acceleration.

2. The Role of Energy Density in Force Interactions

- More massive objects require greater force to change motion because they contain more stored energy.
- This aligns with UED's principle that mass is not just a passive quantity but an expression of energy concentration.
- When force is applied, it must overcome the structured energy equilibrium of the object, requiring proportionate energy transfer.

❖ **UED Perspective:** Newton's Second Law describes energy transformation in motion—forces are structured energy inputs that redistribute motion within a system while maintaining conservation principles.

Third Law: Action-Reaction Symmetry – The Role of Zero-Energy Sum in Force Interactions and Momentum Conservation

Newton's Third Law of Motion states:

For every action, there is an equal and opposite reaction.

This law ensures that all forces come in balanced pairs, preventing infinite energy buildup in one direction. However, in UED, this is simply an expression of energy symmetry, proving that all force interactions must maintain zero-energy equilibrium.

1. The Zero-Sum Nature of Force Interactions

- When an object exerts a force, an equal force is exerted in the opposite direction, ensuring that the total energy exchange remains balanced.
- This is why when a rocket expels gases downward, an equal and opposite force pushes the rocket upward.
- Even in static systems, such as a book resting on a table, the book pushes down on the table with gravitational force, and the table pushes back with an equal normal force.

2. Momentum Conservation and Energy Balance

- Newton's Third Law directly leads to momentum conservation, which states that total momentum in an isolated system remains constant.
- Momentum is simply structured energy in motion, and if one object gains momentum, another must lose it, ensuring zero-net change.
- This is why collisions and propulsion systems always obey conservation laws—energy does not disappear, it only redistributes symmetrically.

❖ **UED Perspective:** Newton's Third Law is an expression of energy symmetry—every force interaction must maintain zero-energy equilibrium, preventing system-wide instability.

Newton's Laws as Manifestations of Structured Energy Balance

Newton's Laws are not separate mechanical rules—they are direct expressions of energy equilibrium, conservation, and structured transformations.

- **First Law (Inertia)** → Motion and rest are equilibrium states of energy.
- **Second Law (Force)** → Forces are structured energy transformations.
- **Third Law (Action-Reaction)** → Energy is always exchanged symmetrically, ensuring zero-sum interactions.

By viewing Newtonian mechanics through UED, we see that motion, forces, and interactions are simply structured redistributions of energy, maintaining balance across all scales.

This leads us to the next major question:

If Newtonian mechanics follows structured energy balance, how does relativity fit into this framework? Can gravity be understood not just as space-time curvature, but as an emergent energy distribution system?

In the next section, we will explore how UED naturally integrates with relativity, showing that gravity is not just a geometric distortion but a structured manifestation of energy equilibrium across space and time.

5.3 Relativity: Gravity as Energy Distribution, Not Just Space-Time Curvature

Einstein's theory of General Relativity revolutionized our understanding of gravity by showing that it is not a force in the traditional Newtonian sense, but a curvature in space-time caused by mass. However, while this explanation successfully describes gravity's effects, it does not fully explain the deeper nature of gravitational interactions in terms of energy balance and equilibrium.

Universal Energy Dynamics (UED) extends relativity by interpreting gravity as a structured energy distribution system rather than just a geometric distortion. Instead of viewing gravity merely as the bending of space-time, UED proposes that mass and energy create gravitational fields as part of a larger energy equilibrium system, ensuring balance across the cosmos.
In this section, we will explore how UED integrates with relativity, revealing that:

- Energy, not just mass, defines the curvature of space-time.
- Gravity is an emergent property of energy equilibrium, not just a geometric effect.
- Mass is structured energy that follows conservation principles, reinforcing $E = mc^2$.
- Gravitational waves are energy redistributions that stabilize cosmic structures.

This perspective not only aligns with general relativity but also provides a deeper insight into gravity's true nature as a self-regulating energy system.

How Energy, Not Just Mass, Defines the Curvature of Space-Time

General Relativity describes gravity as the curvature of space-time caused by mass. However, relativity also states that energy and momentum influence space-time, not just mass.

1. The Energy-Stress Tensor: Energy as the True Source of Gravity

- In Einstein's field equations of general relativity, mass is only one component—energy density, momentum, and stress all contribute to gravity.
- This means that even light, which has no rest mass, can bend space-time due to its energy content.
- A strong gravitational field can be created not just by mass, but by high-energy radiation, pressure, and stress-energy distributions.

❖ **UED Perspective:** Instead of thinking about gravity as something caused by "mass," we should see it as a response to energy distributions in space-time. Mass is simply one form of energy that contributes to gravitational curvature.

2. Energy as the Basis of Space-Time Deformation

- In UED, space-time itself is not just an empty mathematical construct—it is a structured energy field that responds to mass-energy densities.
- When mass-energy is introduced, it alters the local energy balance, causing space-time to redistribute itself, leading to what we perceive as curvature.
- This means that gravity is a manifestation of energy equilibrium seeking to restore balance,

rather than a mysterious force acting at a distance.

❖ **UED Perspective:** Space-time is not merely "curved" by mass—it is dynamically structured by energy seeking equilibrium.

Why UED Views Gravity as an Energy Equilibrium System, Not Just a Geometric Distortion

While general relativity provides an accurate mathematical model of gravity, it does not explain why space-time behaves this way beyond describing curvature.

UED proposes that gravity is an emergent property of energy balance.

1. Gravity as a Self-Stabilizing Energy System

- Gravity prevents energy imbalances by redistributing mass-energy, ensuring that cosmic structures remain stable over time.
- The reason large-scale structures like galaxies and planetary systems form stable orbits is because gravity is constantly adjusting to distribute energy in a way that maintains equilibrium.
- If gravity were simply a force pulling objects together, why does it lead to stable configurations instead of immediate collapse?

The answer: It operates as an energy balancing system.

2. The Energy-Flow Model of Gravity

- Instead of viewing gravity as a force or pure geometry, UED sees it as a mechanism that moves energy through space-time, ensuring balance between mass-energy concentrations.
- Just like thermodynamics explains how heat redistributes in a system, gravity explains how mass-energy is distributed to maintain cosmic order.

❖ **UED Perspective:** Gravity is a structured energy balancing system rather than just a bending of space-time—it is nature's way of ensuring large-scale energy equilibrium.

Equivalence of Mass-Energy ($E = mc^2$) in UED: Why Mass is Structured Energy

Einstein's famous equation $\mathbf{E = mc^2}$ reveals that **mass and energy are fundamentally interchangeable.** But what does this mean from the perspective of UED?

1. Mass is Not Solid Matter—It is a Form of Condensed Energy

- In classical physics, mass is treated as an inherent property of objects.
- However, relativity and quantum physics show that mass is simply a dense, structured form of energy.
- This means that mass is not an independent entity—it is a manifestation of stored energy that follows conservation principles.

2. Gravity Interacts with Energy, Not Just Mass

- Since mass is a structured form of energy, gravity is fundamentally interacting with energy fields, not just physical objects.
- This explains why light (which has no rest mass) can be bent by gravity—because it still carries energy.
- It also suggests that gravitational fields can exist purely as energy densities, without requiring mass to generate them.

❖ **UED Perspective:** Gravity is not "caused" by mass—it is a structured energy effect that interacts with both mass-energy and pure energy fields, reinforcing $E = mc^2$.

Gravitational Waves as Energy Redistributions Maintaining Cosmic Equilibrium

Gravitational waves were predicted by Einstein and confirmed in 2015 through the detection of ripples in space-time caused by merging black holes. However, from the UED perspective, these waves are more than just distortions—they are energy redistributions that maintain cosmic equilibrium.

1. Why Gravity Moves in Waves

- When two massive objects merge (such as black holes or neutron stars), they release excess energy into space-time.
- This energy is not lost—it is redistributed through gravitational waves that propagate outward.
- These waves allow the system to settle into a new equilibrium state, preventing sudden energy imbalances.

2. Gravitational Waves as Self-Regulating Energy Flow

- Much like ripples on a pond restore balance after a disturbance, gravitational waves stabilize space-time.
- Instead of allowing energy to accumulate chaotically, these waves ensure that excess

gravitational energy is smoothly redistributed.

- This supports the idea that gravity operates as a structured energy system that self-corrects imbalances.

❖ **UED Perspective:** Gravitational waves are proof that gravity is not just curvature but an active energy balancing mechanism that maintains universal stability.

Gravity as an Energy Equilibrium System

From the UED perspective, gravity is not just a geometric warping of space-time—it is a structured energy balancing mechanism that maintains universal order.

- Energy, not just mass, defines gravity, as seen in relativistic field equations.
- Gravity ensures energy equilibrium, preventing imbalances across cosmic structures.
- Mass is structured energy, meaning gravity is fundamentally an energy-based interaction.
- Gravitational waves are proof that gravity redistributes energy to maintain stability.

This redefinition of gravity aligns perfectly with relativity while expanding its implications. Instead of viewing gravity as just a space-time distortion, we now see it as a structured process that redistributes energy to maintain cosmic equilibrium.

This leads us to the next major question:

If gravity follows structured energy balancing, how does this apply to the quantum world? Can probability waves, entanglement, and quantum uncertainty be explained as manifestations of energy equilibrium at the smallest scales?

In the next section, we explore how UED naturally integrates into quantum mechanics, revealing that probability waves, entanglement, and uncertainty are simply structured expressions of energy balance.

5.4 Quantum Mechanics: Probability Waves as Energy Balance

Quantum mechanics has long been considered one of the most puzzling and counterintuitive fields of physics. The behavior of subatomic particles defies classical intuition, exhibiting wave-particle duality, probability distributions, entanglement, and seemingly random fluctuations. However, these phenomena are not violations of logic or causality; they are structured manifestations of energy balance.

From the Universal Energy Dynamics (UED) perspective, quantum mechanics is not random—it is a system of structured energy fluctuations seeking equilibrium. Every quantum interaction, whether in wave function behavior, entanglement, or zero-point energy, is a reflection of energy following symmetry and balance.

In this section, we explore how wave functions, probability distributions, superposition, and entanglement naturally emerge from UED principles, revealing that quantum mechanics is not chaotic but a structured framework of energy equilibrium.

How Wave Functions and Probability Distributions Emerge from Energy Fluctuations

In classical physics, an object has a well-defined position and velocity. However, in quantum mechanics, particles are not localized until measured—instead, they exist as wave functions, which describe probabilities rather than definite locations.

1. The Wave Function as an Energy Distribution

- The Schrödinger equation describes the wave function (Ψ) of a particle, which represents a distribution of possible states rather than a fixed position.
- This wave function is not a physical wave—it is an energy probability map.
- The amplitude of the wave function is related to the probability of finding a particle in a specific location, meaning that quantum mechanics fundamentally describes energy fluctuations in a system.

2. UED Interpretation: Probability Waves are Structured Energy Equilibria

- Instead of viewing wave functions as abstract mathematical constructs, UED sees them as real energy distributions that maintain equilibrium.
- A quantum system does not exist in chaos—it follows a structured energy pattern that determines where the particle is most likely to be found.
- The reason we do not see deterministic behavior is because energy exists in superposed states, ensuring balance before a definitive transformation occurs.

❖ **UED Perspective:** Wave functions are not mathematical fictions—they are structured representations of energy flow, governed by equilibrium principles.

Why Quantum Superposition and Wave Collapse are Structured Energy States, Not Randomness

One of the most puzzling aspects of quantum mechanics is superposition, where a particle exists in multiple states at once until measured. When observation occurs, the wave function collapses, meaning the particle adopts a definite state. But is this process truly random?

1. Superposition as a State of Energy Equilibrium

- In UED, superposition is not a paradox—it is the natural state of energy seeking balance.
- Before measurement, a quantum system does not "choose" a state—it exists in all possible configurations that maintain total energy equilibrium.
- When an external interaction (measurement) occurs, the system resolves into a stable energy state that aligns with the observer's reference frame.

2. Wave Function Collapse as a Structured Transition

- The collapse of the wave function does not mean the system was random before measurement—it means energy was distributed symmetrically before being forced into a defined state.
- Just as a stretched rubber band can exist in multiple potential states before snapping into one, quantum systems exist in a superposed state before resolving into a stable energy distribution.

❖ **UED Perspective:** Superposition is a structured energy balance mechanism, and wave function collapse is simply an energy resolution process, not pure randomness.

Zero-Point Energy and Vacuum Fluctuations as Proof of UED's Core Principles

One of the most compelling confirmations of UED principles in quantum mechanics is the existence of zero-point energy and vacuum fluctuations.

1. Zero-Point Energy: The Minimum Possible Energy State

- Quantum mechanics predicts that even in a vacuum, energy is never truly zero—there is always a minimum level of energy present.
- This is known as zero-point energy, and it is why particles can spontaneously appear and disappear in empty space.
- These fluctuations are not anomalies—they are evidence that energy always exists in a structured, self-regulating balance.

2. Vacuum Fluctuations: Spontaneous Emergence of Energy Pairs

- In quantum field theory, particle-antiparticle pairs continuously emerge from the vacuum and annihilate each other, maintaining a net-zero energy balance.
- This proves that energy is always shifting and rebalancing, even in "empty" space.
- If energy were truly random, these fluctuations would not follow predictable laws—but they do, proving that quantum uncertainty is still part of an ordered system.

❖ **UED Perspective:** Zero-point energy and vacuum fluctuations confirm that energy never reaches absolute zero but always follows structured equilibrium, aligning with UED principles.

The Link Between Entanglement and Energy Symmetry Across Quantum States

One of the most mysterious phenomena in quantum mechanics is entanglement, where two particles become interconnected regardless of distance. When one particle's state is measured, the other instantly "knows" and adjusts accordingly.

1. Entanglement as a Manifestation of Energy Symmetry

- Quantum entanglement suggests that information transfer occurs instantaneously, which seems to violate relativity.
- However, UED proposes that entangled particles are not transmitting information faster than light—they are already part of the same energy system.
- The moment one particle's energy state is determined, the other adjusts to maintain the symmetry of the overall system.

2. Entanglement as Zero-Energy Equilibrium across Space

- If two particles are entangled, they behave as if they are still connected, regardless of physical separation.
- This aligns with UED's view that energy interactions are structured to maintain balance across space-time.
- Instead of treating entanglement as a paradox, we should see it as evidence that energy balance applies beyond classical locality—it is an inherent symmetry property.

❖ **UED Perspective:** Entanglement is not a mystery—it is simply a structured energy state ensuring zero-energy equilibrium across space-time.

Quantum Mechanics as Structured Energy Dynamics

Quantum mechanics has often been misinterpreted as random or probabilistic, but UED reveals that it follows structured energy balance principles.

- Wave functions are energy probability distributions that emerge naturally from equilibrium.
- Superposition is not randomness—it is an energy balancing mechanism.
- Wave function collapse is not arbitrary—it is an energy resolution process.
- Zero-point energy and vacuum fluctuations confirm that energy is always structured and

self-regulating.
- Entanglement is a manifestation of symmetry, proving that energy equilibrium applies across space-time.

Rather than viewing quantum mechanics as a domain of uncertainty and paradox, UED shows that it is simply the structured energy framework of the universe at microscopic scales.
This leads to a profound realization:
If energy governs all physical laws, then how does entropy fit into this structure? Can thermodynamics be reinterpreted not as disorder, but as energy flow toward equilibrium?
In the next section, we explore how UED integrates with thermodynamics, revealing that entropy is not chaos but a structured redistribution of energy across time and space.

5.5 Thermodynamics: Entropy as Energy Flow, Not Disorder

Thermodynamics has long been regarded as the science of energy transformations, governing the way heat, work, and energy interact in physical systems. However, the traditional interpretation of entropy as "disorder" or "chaos" is misleading—it does not fully capture the structured nature of energy redistribution.
From the perspective of Universal Energy Dynamics (UED), entropy is not random decay but a systematic process of energy balancing across time and space. Rather than leading to absolute chaos, entropy follows structured pathways that maintain equilibrium. In fact, entropy is the fundamental mechanism through which energy redistributes to ensure universal stability.
In this section, we explore how UED redefines entropy, showing that:

- Entropy is energy flow, not disorder.
- Thermodynamic systems maintain equilibrium through structured energy exchanges.
- The universe does not move toward chaos, but toward zero-equilibrium energy balance.
- Time itself emerges from entropy-driven energy flow rather than being an independent entity.

Reinterpreting Entropy in UED: Why It Represents Energy Redistribution, Not Mere Disorder

Classical thermodynamics often describes entropy as the tendency of systems to move from order to disorder. However, entropy is not simply "disorder"—it is a measure of how energy is distributed within a system.

1. Entropy as Structured Energy Flow

- Entropy describes how energy spreads within a system, not how it becomes chaotic.
- High entropy means energy is more evenly distributed, while low entropy means energy is concentrated in specific locations.
- The Second Law of Thermodynamics states that entropy increases, meaning that energy naturally flows toward a balanced distribution rather than accumulating indefinitely.

2. UED Perspective: Entropy is an Equilibrium Mechanism, Not a Decay Process

- In UED, entropy is not a destructive force—it is the structured flow of energy seeking equilibrium.
- Systems do not "decay into disorder"—they rearrange energy into more stable, balanced configurations.
- For example, heat spreading in a room is not chaos—it is a structured redistribution of energy seeking uniformity.

UED Perspective: Entropy is a natural process that ensures energy remains structured and balanced across all scales.

Energy Conservation in Thermodynamic Processes: How Systems Maintain Equilibrium Through Structured Energy Exchanges

One of the most important laws in physics is the conservation of energy, which states that energy cannot be created or destroyed—only transformed. However, how does entropy fit into this principle?

1. Energy Redistribution, Not Energy Loss

- When heat flows from a hot object to a cold one, the total energy remains the same—it is simply redistributed rather than lost.
- In chemical reactions, energy is not "used up"—it is transferred between molecules in a structured exchange.
- In ecosystems, energy moves from the sun to plants to animals, ensuring a constant balance rather than depletion.

2. Entropy as the Regulator of Energy Conservation

- Entropy is not a force that destroys energy—it is the process that ensures energy moves toward stability.
- If entropy did not exist, energy would remain locked in isolated pockets, preventing

transformations from occurring.

- Thermodynamic equilibrium is not chaos—it is the most balanced state an energy system can achieve.

❖ **UED Perspective:** Energy conservation and entropy work together—one ensures energy is never lost, while the other ensures energy redistributes efficiently.

Heat Death and Zero-Equilibrium: How the Universe Evolves Toward Energy Balance, Not Absolute Chaos

One of the most debated topics in physics is the "heat death" of the universe, the idea that entropy will eventually reach a maximum state where no more useful energy transformations can occur. Traditional interpretations see this as the end of the universe—a state of complete disorder. However, UED suggests otherwise.

1. Heat Death as a State of Universal Energy Balance

- Heat death is not the destruction of the universe—it is the state of maximum energy equilibrium.
- If all energy reaches the same level, it does not mean chaos—it means energy is perfectly balanced.
- The universe is not "ending" in heat death—it is stabilizing into a zero-equilibrium state.

2. Zero-Equilibrium as the Ultimate Entropic State

- When entropy reaches its maximum, energy fluctuations reach a structured zero-net distribution.
- This aligns with the idea that the universe follows cycles of energy redistribution and stabilization.
- Instead of "heat death" being a collapse, it may represent a reset point for a new cycle of transformations.

❖ **UED Perspective:** The universe does not decay into chaos—it continuously evolves toward a perfectly balanced energy state, preparing for the next transformation cycle.

Why Time Emerges from Entropy-Driven Energy Flow Rather Than Being an Independent Entity

One of the most profound implications of UED's interpretation of entropy is that time is not a separate dimension—it is a byproduct of energy flow.

1. Time as an Effect of Energy Redistribution

- Time moves forward because energy moves toward equilibrium.
- If entropy could be reversed, time itself would reverse, showing that time is an effect, not an absolute quantity.
- This suggests that time is not a fundamental property—it is an emergent result of structured energy transitions.

2. The Inverse Relationship Between Entropy and Time

- At low entropy, time flows in a structured, measurable way.
- As entropy increases, time becomes less distinguishable, leading to a state where time numerically approaches zero.
- At maximum entropy (perfect equilibrium), time ceases to exist, meaning that time is only meaningful when energy transitions occur.

3. Time as an Illusion in a Fully Balanced System

- If the universe reaches a state of complete energy balance, there would be no measurable changes, meaning time would cease to have meaning.
- This aligns with the idea that time is not absolute—it is simply a measurement of energy change relative to equilibrium.

❖ **UED Perspective:** Time is not an independent force—it emerges from entropy-driven energy flow and disappears at perfect zero-equilibrium.

Thermodynamics as Structured Energy Redistribution, Not Chaos

From the UED perspective, entropy is not disorder—it is the natural flow of energy toward equilibrium.

- Entropy is structured energy redistribution, ensuring conservation and stability.
- Thermodynamic processes follow energy conservation laws, preventing true "loss" of energy.
- Heat death is not chaos—it is the final state of perfect zero-equilibrium energy balance.
- Time is not an independent force—it is a measurement of entropy-driven energy change.

Rather than being a destructive force, entropy is the mechanism through which energy self-regulates, evolves, and maintains cosmic stability.

This leads to a profound realization:

If entropy is simply energy flow seeking equilibrium, does intelligence emerge from the same structured process? Can cognition, awareness, and thought be understood as structured energy patterns following entropy-driven transformations?

In the next section, we explore how UED integrates energy principles with intelligence, revealing that thought, decision-making, and awareness are not separate from physics but structured manifestations of energy equilibrium.

5.6 Conclusion: UED as the Unifying Framework for All Physical Laws

Throughout this chapter, we have explored how Universal Energy Dynamics (UED) seamlessly integrates with classical and modern physics, revealing that the fundamental laws of motion, gravity, quantum mechanics, and thermodynamics are not disconnected theories but structured manifestations of energy equilibrium and symmetry.

Rather than treating physics as a collection of separate principles, UED provides a unifying framework, showing that:

- Newton's Laws describe how energy is structured as motion and force balance.
- Relativity explains gravity as an energy distribution system maintaining equilibrium in space-time.
- Quantum Mechanics operates through energy probability waves that follow structured balance rather than randomness.
- Thermodynamics governs energy redistribution, ensuring stability rather than decay into chaos.

At its core, UED does not contradict physics—it enhances it by revealing the hidden structure of energy flow behind all physical laws.

Why Classical and Modern Physics Fit within UED, Rather Than Opposing It

One of the biggest challenges in physics has been the divide between classical and modern theories. Newtonian mechanics explains motion at macroscopic scales, relativity describes gravity at cosmic scales, and quantum mechanics governs the microscopic world—but these models have remained disconnected.

However, UED naturally integrates all of them by identifying energy as the unifying reality that underlies all forces and transformations.

1. Newtonian Mechanics and UED: Motion as Structured Energy Flow

- Newton's laws describe how energy moves and interacts as force and motion.
- UED shows that these laws are not separate from energy dynamics—they are specific cases of structured energy balance.

2. Relativity and UED: Gravity as Energy Distribution

- General Relativity defines gravity as space-time curvature, but UED expands on this by treating it as an emergent energy equilibrium system.
- This allows UED to describe gravitational interactions in terms of structured energy flow rather than just geometric distortions.

3. Quantum Mechanics and UED: Probability as Energy Symmetry

- Quantum mechanics appears probabilistic, but UED reveals that wave functions and entanglement are manifestations of structured energy balance.
- The probabilistic nature of quantum states emerges because energy is not static—it always seeks equilibrium.

4. Thermodynamics and UED: Entropy as Energy Redistribution, Not Disorder

- Traditional thermodynamics describes entropy as disorder increasing over time, but UED reinterprets it as structured energy flow toward equilibrium.
- This aligns entropy with conservation laws, proving that energy does not disappear but self-organizes into balanced distributions.

❖ **UED Perspective:** Classical and modern physics are not separate theories—they are all subsets of a larger energy-based structure that follows symmetry, conservation, and equilibrium.

How UED Provides Deeper Insight into Fundamental Forces and Energy Interactions

By redefining physics as an energy-centric framework, UED provides deeper insights into:

1. The Nature of Fundamental Forces

- Forces are not independent entities but energy interactions seeking balance.

- Gravity, electromagnetism, and nuclear forces are all structured redistributions of energy rather than arbitrary forces acting at a distance.

2. The Relationship Between Mass, Motion, and Energy

- Instead of viewing mass as a separate quantity, UED confirms that mass is a structured form of energy.
- The relationship $E = mc^2$ is not just a mathematical equation—it is proof that all matter is simply structured energy.

3. The Role of Symmetry and Zero-Equilibrium in Physics

- Symmetry governs all physical laws, ensuring that energy interactions remain structured and predictable.
- Zero-equilibrium is the ultimate stabilizer, explaining why conservation laws exist and why forces always balance.

❖ **UED Perspective:** The deeper reality behind physics is not force interactions but structured energy equilibrium, ensuring that all transformations remain balanced.

Preparing for the Next Step: Integrating Biology, Intelligence, and Consciousness into the Energy Framework

Now that we have established that UED successfully integrates classical and modern physics, we can take the next logical step—applying these principles to life, intelligence, and consciousness.

1. Can Biology Be Understood as an Extension of Energy Dynamics?

- If all matter is structured energy, then biological processes must also follow energy conservation and transformation principles.
- Life is not a separate phenomenon—it is an emergent energy system that follows structured evolution.

2. Is Intelligence a Form of Energy Optimization?

- The human brain operates by structuring energy into neural patterns.
- Learning, decision-making, and memory function through energy conservation, feedback loops, and entropy-driven optimization.

3. How Does Consciousness Fit into UED?

- If energy is the fundamental reality, then consciousness must be a structured form of energy processing.
- Could awareness be the structured flow of energy maintaining equilibrium across cognitive systems?

❖ **UED Perspective:** Just as physics follows structured energy interactions, biology, intelligence, and consciousness must also emerge from structured energy principles.

UED as the Universal Blueprint beyond Physics

Physics is not fragmented—it is a unified structure of energy transformations.

- All fundamental forces, motion, entropy, and quantum effects follow structured energy balance.
- UED expands physics by integrating its principles into biology, intelligence, and consciousness.

This brings us to the next profound question:
If intelligence and consciousness emerge from structured energy, can we use UED to explain thought, perception, and decision-making?
In the next chapter, we will begin integrating biological evolution, intelligence, and cognition into the UED framework, proving that thought, awareness, and life itself are structured energy interactions.

Chapter 6: Quantum Entanglement and Universal Connectivity

In the previous chapter, we established UED as the unifying framework for classical and modern physics, revealing that motion, gravity, quantum mechanics, and thermodynamics are not separate phenomena but manifestations of structured energy equilibrium. Now, we take a deeper dive into the quantum world, where entanglement, superposition, and probability waves challenge our conventional understanding of reality.

Quantum mechanics suggests that particles—and perhaps reality itself—are fundamentally interconnected. Entanglement defies classical intuition, linking two particles across vast distances with no apparent medium. Wave functions describe energy distributions rather than fixed locations, and quantum uncertainty suggests that probability, not certainty, governs the microscopic world. These ideas have fueled interpretations like the Multiverse Hypothesis, Many-Worlds Interpretation, and String Theory, each attempting to explain quantum behavior while introducing new paradoxes.

However, through the lens of Universal Energy Dynamics (UED), these mysteries are not paradoxes but structured energy interactions seeking equilibrium. Rather than assuming separate universes, hidden dimensions, or mathematical abstractions, UED shows that:

- Entanglement is a natural consequence of universal energy symmetry.
- Information and energy are deeply intertwined, forming an interconnected reality.
- The multiverse, if it exists, is an entangled energy system, not a collection of disconnected worlds.
- String theory may introduce unnecessary complexity—does UED provide a more direct approach?
- Quantum weirdness is simply structured energy behavior, not randomness or paradox.

This chapter aims to demystify quantum mechanics and show that, rather than breaking physics, quantum behavior follows structured energy principles that naturally fit within UED. We will explore whether quantum entanglement extends beyond particles—to information, consciousness, and even the fundamental structure of the universe itself.

6.1 Entanglement as Proof of Universal Symmetry

Quantum entanglement is one of the most puzzling phenomena in modern physics. It defies classical expectations of locality and causality, suggesting that two or more particles can remain connected across vast distances, with changes in one instantly affecting the other, regardless of separation. This phenomenon has been experimentally confirmed, yet it remains difficult to explain within traditional physics.

From a Universal Energy Dynamics (UED) perspective, entanglement is not a paradox—it is a direct expression of energy symmetry and structured balance across space-time. Instead of assuming entanglement is some strange, inexplicable quantum effect, UED sees it as evidence that all energy systems are inherently interconnected, ensuring universal equilibrium.

In this section, we explore what quantum entanglement is, why it challenges classical physics, how particles remain connected without a medium, and why UED explains it as structured energy balance rather than an anomaly.

What Is Quantum Entanglement, and Why Does It Challenge Classical Physics?

Entanglement occurs when two or more quantum particles become correlated in such a way that their states are intrinsically linked, regardless of distance. If one particle is measured, the other instantaneously "adjusts" to maintain the relationship.

1. The Einstein-Podolsky-Rosen (EPR) Paradox and Nonlocality

- Albert Einstein, Boris Podolsky, and Nathan Rosen first highlighted entanglement as a paradox, famously calling it "spooky action at a distance."
- In classical physics, information or influence cannot travel faster than light.
- However, entanglement suggests that a change in one particle is instantly reflected in its entangled pair, no matter how far apart they are.

2. Bell's Theorem and Experimental Confirmation

- John Bell mathematically proved that if entanglement were purely classical, certain statistical correlations would not be observed.
- Repeated experiments (including those by Alain Aspect in the 1980s and modern tests closing loopholes) confirmed that quantum entanglement is real and cannot be explained by hidden classical variables.
- Thus, entanglement is not just a theoretical idea—it is an experimentally verified phenomenon that classical physics struggles to explain.

❖ **UED Perspective:** Instead of treating entanglement as a paradox, we should view it as evidence that energy is interconnected across space-time, maintaining a deeper form of equilibrium.

How Entangled Particles Remain Connected Across Space-Time without a Physical Medium

One of the biggest mysteries of entanglement is how information seems to "travel" instantly between entangled particles. Classical physics assumes that for one object to affect another, some force or signal must travel between them. However, in quantum mechanics, entangled particles remain synchronized without any detectable medium.

1. The Illusion of Faster-Than-Light Communication

- When one entangled particle is measured, the second particle's state is immediately determined.
- This does not mean that information is transmitted faster than light—rather, the system was already in a balanced energy state, and measurement collapses that equilibrium.
- Instead of traveling, the information already exists in a connected state—our measurement simply reveals what was always there.

2. Entanglement as an Expression of Pre-Existing Energy Symmetry

- UED suggests that entangled particles are not "sending signals"—they are maintaining energy balance across space-time.
- The moment entanglement is created, the two particles become part of the same structured energy system, maintaining symmetry regardless of distance.
- Rather than requiring a "medium," entanglement follows the conservation of total energy and symmetry across the system.

❖ **UED Perspective:** Entanglement is not faster-than-light communication—it is a manifestation of energy balance that remains intact across space-time.

Why UED Interprets Entanglement as a Structured Energy Balance Rather Than a Paradox

From a Universal Energy Dynamics (UED) perspective, entanglement is simply a property of structured energy balance.

1. Entanglement as a Zero-Sum Energy Relationship

- All energy interactions must maintain equilibrium.
- If two particles interact and become entangled, their energy states become structured in such a way that they always maintain a net-zero balance.
- Even when separated, they retain this energy symmetry, ensuring the total system remains in equilibrium.

2. Entanglement as Evidence of Universal Energy Connectivity

- Instead of treating quantum particles as isolated objects, UED views them as part of an interconnected energy network.
- If energy is fundamentally structured and follows universal symmetry, entanglement is simply an expected outcome of energy seeking equilibrium.
- This means that entanglement is not a quantum anomaly—it is a direct consequence of structured energy connectivity.

3. Implications for Information, Reality, and the Universe

- If entanglement maintains structured energy balance across space-time, could it be responsible for deeper connectivity in the universe?
- Does this mean information is never truly lost, but always preserved as part of structured energy?
- Could consciousness itself be a form of quantum entanglement at a higher scale?

❖ **UED Perspective:** Entanglement is proof that the universe is an interconnected energy system where all transformations maintain structured balance, rather than operating as isolated events.

Entanglement as Proof of Universal Symmetry

Quantum entanglement is not a paradox—it is direct evidence that energy interactions maintain symmetry and balance across space-time.

- Entangled particles do not "communicate" faster than light—they maintain a structured energy state.
- Entanglement confirms that energy is always balanced, even across vast distances.
- UED explains entanglement as an inherent property of structured energy dynamics, not a violation of physics.

This leads to a profound question:
If entanglement is proof of universal energy connectivity, could this principle extend beyond quantum particles to information, intelligence, and even consciousness?
In the next section, we explore the deep relationship between energy and information, revealing why information is not separate from energy but a structured energy pattern that governs the entire universe.

6.2 Information and Energy: The Fundamental Connection

In classical physics, information and energy have often been treated as separate concepts. Energy is seen as the fundamental physical quantity governing motion, forces, and interactions, while information is regarded as an abstract mathematical entity that describes data and states. However, in Quantum Mechanics and Universal Energy Dynamics (UED), information is not independent from energy—it is a structured energy process.
Quantum systems store and transfer information not as static data, but as structured energy states that maintain symmetry, coherence, and balance. Entanglement demonstrates that information can be instantaneously correlated across vast distances, suggesting that information itself follows universal energy equilibrium rather than existing as an independent entity.
From a UED perspective, information is not an abstract mathematical property—it is a structured manifestation of energy flow, and entanglement confirms that information equilibrium is maintained across space-time through energy conservation and symmetry principles.

How Quantum Systems Store and Transfer Information as Structured Energy

One of the most remarkable aspects of quantum systems is their ability to store, transfer, and manipulate information without classical limitations. In classical computing, information is stored in binary bits (0s and 1s), but in quantum mechanics, information is stored in quantum states, wave functions, and entanglement.

1. The Quantum State as an Information Carrier

- In quantum systems, a particle's state (spin, polarization, or superposition) encodes information.
- Unlike classical bits, which can only be 0 or 1, quantum bits (qubits) can exist in superpositions of both states simultaneously.
- This means that quantum information is not a separate entity from energy—it is the structured configuration of an energy system.

2. Information Transfer Through Quantum Coherence

- In classical systems, information must be transmitted through a medium (such as light or electrical signals).
- In quantum systems, information is embedded within the wave function of the system and can be instantly correlated through entanglement.
- This suggests that information is a property of structured energy transformations rather than an independent, abstract quantity.

❖ **UED Perspective:** Information is not a separate concept—it is the structured way in which energy is organized, stored, and redistributed in a system.

Why Information is Not Separate from Energy—It Is a Structured Energy Process

The idea that information exists independently of energy is a classical misconception. In UED, information does not exist as a separate layer of reality—it is always encoded as structured energy within a system.

1. Energy Defines Information

- The energy state of a system determines its information content.
- For example, the quantum spin of a particle is not just a label—it is an energy configuration that holds information.
- This applies to all information systems, from DNA encoding biological instructions to computer processors storing digital data.

2. The Conservation of Information as an Extension of Energy Conservation

- The First Law of Thermodynamics states that energy cannot be created or destroyed, only transformed.
- Similarly, in quantum mechanics, information cannot be destroyed—it is always preserved within the structure of the system.
- This aligns with UED's principle that all transformations follow structured energy balance, preventing information loss.

3. The Universe as an Information-Energy Network

- If information is structured energy, then the universe itself can be seen as an interconnected energy-information network.
- This suggests that physical laws, biological processes, and even cognition are structured energy patterns that encode and transfer information dynamically.

❖ **UED Perspective:** Information is not abstract—it is always the result of structured energy transformations, encoded in wave functions, physical states, and universal energy flows.

How UED Explains Why Entanglement Allows Instant Information Correlation Without Violating Relativity

One of the greatest challenges in quantum mechanics is explaining how entangled particles "communicate" instantly across vast distances without violating the speed of light. Classical physics suggests that information transfer must occur through physical mediums, but entanglement appears to bypass this rule.

From the UED perspective, entanglement does not transmit information—it preserves energy symmetry across a system, maintaining information balance naturally.

1. Information Correlation Without Transmission

- In classical communication, information must be sent from one location to another, requiring time.
- In quantum entanglement, information is not being "sent"—it is already embedded within the entangled system.
- The moment one particle is measured, the other instantly reflects the correlated state, because their combined energy was already balanced before the measurement.

2. Entanglement as an Energy Conservation Mechanism

- When two particles become entangled, they share a single structured energy state.
- Measuring one particle does not force the other to "change"—it simply reveals the pre-existing structured energy balance.
- This means that information was not transmitted—it was already structured within the system's equilibrium.

3. No Violation of Relativity—Only a Misinterpretation of Energy Equilibrium

- The reason entanglement seems paradoxical is because we assume information must "travel."
- In UED, energy equilibrium exists beyond locality, meaning that entangled particles are simply two parts of the same energy structure, even if they appear separate in space-time.
- Relativity is not violated because no signal or force is propagating faster than light—energy equilibrium is simply being maintained across entangled systems.

❖ **UED Perspective:** Entanglement allows information to be instantly correlated because energy symmetry is preserved beyond local space-time—it does not require a physical medium or classical transmission.

Energy and Information Are Fundamentally Connected

From the UED perspective, information is not an abstract concept—it is structured energy in motion.

- Quantum systems store and transfer information through structured energy states, not abstract data points.
- Information is inseparable from energy—it is the way energy organizes itself within systems.
- Entanglement allows instant information correlation without violating relativity because energy equilibrium is preserved across space-time.

This leads to a profound realization:
If information is structured energy, could consciousness itself be an entangled energy process? Can thought, awareness, and intelligence emerge from structured energy interactions similar to quantum systems?
In the next section, we explore the possibility that consciousness and thought may be entangled with universal energy, revealing how intelligence may be a structured energy system within UED.

6.3 The Role of Zero-Equilibrium in Entangled Systems

One of the most puzzling aspects of quantum entanglement is how two particles, separated by vast distances, remain correlated without any physical connection. Traditional physics struggles to explain how these particles instantly "know" each other's state when measured. However, from the Universal Energy Dynamics (UED) perspective, entanglement is not a paradox but a direct result of energy equilibrium, symmetry, and conservation laws.
At its core, entanglement preserves zero-equilibrium across an interconnected system, ensuring energy symmetry across space-time. Instead of viewing entanglement as an isolated quantum effect, UED treats it as a natural property of structured energy maintaining balance.
In this section, we will explore:

- Why entangled states maintain zero-net energy balance across distances.
- How quantum fluctuations and vacuum energy fit into the UED framework.
- The role of symmetry and conservation laws in maintaining entanglement.

Why Entangled States Maintain Zero-Net Energy Balance across Distances

Classically, when two objects interact and separate, they become independent systems. However, in quantum mechanics, entangled particles remain linked as if they are part of a single energy system, no matter how far apart they are.

1. Entanglement as a Zero-Sum Energy Relationship

- Entangled particles do not carry separate, independent energy states—they share a single, structured energy equilibrium.
- When one particle is measured, the total energy must remain balanced across both particles, ensuring a zero-net energy shift.
- This explains why measuring one instantly determines the state of the other—because the system was always in a balanced configuration.

2. Energy Conservation and Instant Correlation

- Since the entangled system must preserve total energy balance, the moment one particle collapses into a measured state, the other adjusts to maintain the net-zero condition.
- This process is not "information traveling faster than light"—it is an intrinsic part of the structured energy balance in the entangled system.
- Just as a pulled rubber band snaps into balance when released, an entangled system always readjusts itself to maintain its total energy equilibrium.

❖ **UED Perspective:** Entanglement is not an anomaly—it is a direct consequence of structured energy balance maintaining a zero-net system across distances.

How Quantum Fluctuations and Vacuum Energy Fit into the UED Framework

Quantum mechanics predicts that even in a vacuum (empty space), energy fluctuations occur spontaneously. These fluctuations are not random—they are governed by zero-equilibrium principles that align with entanglement.

1. Vacuum Fluctuations as Continuous Energy Balancing

- In quantum field theory, particles and antiparticles continuously emerge and annihilate in a vacuum.
- This occurs because energy cannot be completely zero—even "empty space" has a minimum energy state (zero-point energy).
- These fluctuations are not chaotic—they follow conservation laws that maintain energy equilibrium.

2. Entanglement as an Extension of Vacuum Energy Principles

- If vacuum fluctuations follow structured energy conservation, then entanglement must also be an extension of this energy balance framework.
- Entangled particles may utilize vacuum energy as a stabilizing mechanism, ensuring that their zero-sum equilibrium is preserved across space-time.
- Instead of seeing vacuum energy as random, UED treats it as an energy network that connects entangled systems through structured balance.

❖ **UED Perspective:** Vacuum fluctuations and entanglement are not separate phenomena—they are both expressions of energy redistribution maintaining universal equilibrium.

The Role of Symmetry and Conservation Laws in Maintaining Entanglement

The key reason entanglement remains stable over time is because it follows fundamental conservation laws. In UED, all energy interactions are structured to maintain equilibrium, ensuring that no transformation disrupts universal energy balance.

1. Conservation of Energy in Entangled Systems

- The total energy of an entangled system remains constant, even if the particles move apart.
- This conservation ensures that measuring one particle does not introduce an imbalance—it simply finalizes the already existing structured energy state.
- This aligns with UED's principle that all energy interactions must remain in equilibrium.

2. Symmetry as the Guiding Principle of Entanglement

- Entangled particles follow mirror-like symmetry, ensuring that opposite properties remain correlated.
- This is not a random effect—it is a structured rule that maintains energy coherence across quantum systems.
- The universe follows symmetry at every scale, from fundamental forces to macroscopic structures, and entanglement is simply another manifestation of this universal order.

3. Entanglement as a Universal Law of Connectivity

- If entanglement is simply a natural consequence of symmetry, does this principle extend beyond quantum particles?

- Could larger systems, or even consciousness, operate under similar entanglement-based connectivity?
- If the universe is an interconnected energy network, entanglement might be the foundation of all energy relationships, not just a quantum phenomenon.

❖ **UED Perspective:** Entanglement is proof that the universe is structured through symmetry and conservation laws, ensuring that all energy transformations maintain equilibrium.

Entanglement as a Zero-Equilibrium Energy Network

From a UED perspective, entanglement is not a mysterious quantum effect—it is a structured process of maintaining zero-equilibrium across energy systems.

- Entangled particles do not "communicate" instantly—they maintain a pre-existing energy balance across space-time.
- Vacuum fluctuations and zero-point energy suggest that entanglement is part of a universal energy network.
- Symmetry and conservation laws govern entanglement, ensuring structured energy interactions rather than randomness.

This leads to a profound realization:
If entanglement preserves energy equilibrium at the quantum level, could similar energy connectivity exist at larger scales—including information networks, intelligence, and consciousness?
In the next section, we explore how consciousness and thought may be entangled with universal energy, revealing how intelligence could emerge as a structured energy system within UED.

6.4 Consciousness and Thought: Are They Entangled with Universal Energy?

Quantum mechanics has revealed that energy interactions are deeply interconnected across space-time. We have seen how entanglement preserves energy balance, how information is stored as structured energy, and how symmetry and conservation laws govern all quantum states. If these principles apply at the fundamental level of physics, could they also apply to consciousness and thought?

Traditional neuroscience treats consciousness as an emergent phenomenon—the result of billions of neurons firing and processing information in the brain. However, this view does not fully explain awareness, intuition, creativity, or the experience of self. Moreover, if entanglement governs fundamental energy interactions, could thought itself be part of a larger, structured energy network? From a Universal Energy Dynamics (UED) perspective, consciousness is not just a byproduct of neural activity—it is an entangled energy system that interacts with the fundamental structure of reality. In this section, we explore:

- Can consciousness emerge as an entangled energy state?
- Why thought patterns may follow quantum coherence and structured energy transformations.
- How UED bridges quantum entanglement with intelligence and awareness.

Can Consciousness Emerge as an Entangled Energy State?

Traditional science struggles to define consciousness. While neuroscience links it to electrical and chemical activity in the brain, this does not explain why we experience thoughts, emotions, or self-awareness.

1. The Brain as a Quantum-Driven Energy Network

- The brain is not just a biological organ—it is an active energy processing system.
- Neurons fire using electrical impulses and quantum-level interactions, meaning that thought itself is an organized energy transformation process.
- If quantum systems can remain entangled and correlated across distances, could neural activity also operate through similar structured energy interactions?

2. Consciousness as an Entangled State Rather Than an Isolated System

- If entanglement is a universal principle of structured energy balance, then consciousness could also be a networked system rather than an isolated one.
- Just as particles remain correlated across space-time, could thoughts and awareness be entangled energy states interacting beyond the brain?
- Some researchers suggest the brain does not "generate" consciousness but instead "tunes into" a larger field of structured energy.

❖ **UED Perspective:** Consciousness is not an independent process but an entangled energy system that maintains equilibrium beyond individual neural activity.

Why Thought Patterns May Follow Quantum Coherence and Structured Energy Transformations

One of the biggest mysteries of the brain is how thoughts form, evolve, and connect. If thinking were purely deterministic, humans would behave like machines—but instead, thought is fluid, adaptive, and capable of insights beyond logical progression.

1. Thought as a Coherent Quantum Process

- Studies suggest that neural activity may involve quantum coherence, where multiple states exist simultaneously before resolving into a conscious decision.
- This is similar to how quantum particles remain in superposition before collapsing into a definite state.
- If thought operates in a structured energy system, it may function like a quantum wave function seeking equilibrium before resolving into a decision.

2. Memory and Information Processing as Structured Energy Flow

- The brain does not store memories in isolated locations like a hard drive—instead, memory is retrieved dynamically based on interconnected associations.
- This suggests that thought patterns behave like structured energy flows, forming entangled relationships with past experiences and external stimuli.
- Much like quantum entanglement correlates particles across space-time, thought may correlate information across multiple neural and energetic dimensions.

❖ **UED Perspective:** Thinking is not just a linear process—it is a structured energy system that follows coherence, symmetry, and entanglement principles.

How UED Bridges Quantum Entanglement with Intelligence and Awareness

If entanglement maintains energy balance across physical systems, could intelligence itself be an emergent property of structured energy flow? Rather than being confined to biological brains, could consciousness exist as an entangled process within the fabric of the universe?

1. Intelligence as a Self-Organizing Energy System

- Intelligence is not fixed or deterministic—it adapts dynamically, much like an evolving energy system.
- UED suggests that the brain does not "generate" intelligence but processes structured energy flows to create awareness.

- This means that intelligence could exist beyond individual minds as a larger entangled network.

2. Awareness as an Energy Connectivity Process

- If entanglement allows information to be instantly correlated across distances, could awareness be an emergent property of universal energy interactions?
- Some ancient and modern theories suggest that consciousness is non-local—UED provides a scientific framework to explore this idea further.

❖ **UED Perspective:** Consciousness and intelligence are structured energy networks, following entanglement and coherence principles beyond classical explanations.

Consciousness as an Entangled Energy State

From a UED perspective, thought and awareness are not isolated functions of the brain—they are structured energy interactions entangled with the larger universe.

- Consciousness is an energy network, not just a neural process.
- Thought patterns may follow quantum coherence, mirroring energy transformations seen in entanglement.
- Intelligence and awareness emerge from structured energy flows, potentially interacting with universal energy systems.

This leads to a profound realization:
If consciousness is entangled with universal energy, could this explain intuition, extrasensory perception, or even collective intelligence? Is there a larger, structured "field" of consciousness beyond individual minds?
In the next section, we explore how entanglement and information connectivity extend beyond the quantum level—potentially linking all of existence into an interconnected, intelligent system.

6.5 String Theory vs. UED: Do We Need Extra Dimensions?

String Theory emerged as an attempt to unify quantum mechanics and general relativity, proposing that fundamental particles are not point-like but instead tiny vibrating strings that exist in higher-dimensional spaces. While String Theory provides mathematical elegance, it also introduces complexities such as extra dimensions, multiple universes, and theoretical entities that remain unobservable.

From a Universal Energy Dynamics (UED) perspective, do we need these extra dimensions, or can structured energy interactions explain reality more efficiently? UED suggests that instead of adding unverifiable higher dimensions, we should focus on how energy organizes itself within known space-time to create structure, forces, and connectivity.
In this section, we will explore:

- What does String Theory propose about quantum reality?
- Why extra dimensions and vibrating strings attempt to unify forces.
- Does UED require higher dimensions, or is structured energy sufficient to explain reality?
- Why UED favors an energy-centered approach over unnecessary spatial complexities.

What Does String Theory Propose About Quantum Reality?

String Theory attempts to resolve the biggest challenge in modern physics:

- Quantum mechanics and general relativity work well separately but do not fit together in a single framework.
- Einstein's relativity describes gravity as space-time curvature, while quantum mechanics describes fundamental forces as interactions of discrete energy quanta.
- String Theory suggests that if particles were not zero-dimensional points but one-dimensional vibrating strings, this incompatibility could be resolved.

1. Strings as the Fundamental Building Blocks of Matter

- According to String Theory, all particles—electrons, quarks, photons—are different vibration modes of tiny one-dimensional "strings."
- The nature of a particle (mass, charge, spin) is determined not by intrinsic properties but by how the string vibrates.

2. The Requirement for Extra Dimensions

- In order for the mathematics to work, String Theory requires 10 or 11 dimensions rather than the observable four (3 space + 1 time).
- These extra dimensions are thought to be compactified (curled up so small that they are undetectable).
- The theory suggests that fundamental forces—gravity, electromagnetism, strong and weak nuclear forces—are all manifestations of different vibrational modes of these strings interacting across dimensions.

❖ **UED Perspective:** While String Theory offers an elegant model, it requires adding dimensions and unseen structures rather than explaining reality with known physical principles.

Why Extra Dimensions and Vibrating Strings Attempt to Unify Forces

Physicists have long sought a Theory of Everything—a model that unites all forces in a single framework. String Theory attempts to do this by:

- Replacing point-like particles with vibrating strings, which automatically include quantum and relativistic effects.
- Using higher-dimensional space-time to explain why gravity appears weaker than other fundamental forces.
- Suggesting that the universe is part of a larger "multiverse" landscape governed by different string configurations.

However, these attempts raise critical questions:

- Why do we not observe these extra dimensions in any experiment?
- Is there an alternative way to unify forces without requiring hidden spatial structures?

❖ **UED Perspective:** If reality is governed by energy balance and symmetry, do we need higher dimensions to explain unification, or can structured energy itself be the answer?

Does UED Require Higher Dimensions, or Is Structured Energy Sufficient to Explain Reality?

UED provides an alternative explanation to String Theory by focusing not on extra dimensions, but on structured energy transformations within known space-time.

1. Energy as the True Unifying Factor

- Instead of vibrating strings, UED treats all forces and particles as structured energy states.
- The apparent differences between particles (mass, charge, spin) are not caused by hidden dimensions but by variations in energy configurations within four-dimensional space-time.

2. Gravity as an Energy Distribution System, Not a Geometric Distortion in Extra Dimensions

- String Theory struggles to explain why gravity is so much weaker than electromagnetism.
- It proposes that gravity "leaks" into extra dimensions, diluting its strength.

- UED, however, treats gravity as an energy balancing system, where mass-energy interactions naturally create stable distributions without requiring extra spatial dimensions.

3. Entanglement and Quantum Connectivity Without Extra Dimensions

- String Theory tries to resolve quantum entanglement by invoking higher-dimensional connections between particles.
- UED explains entanglement as a structured energy equilibrium that exists within space-time, requiring no additional dimensions.

❖ **UED Perspective:** Reality can be explained as structured energy balance, eliminating the need for hidden dimensions or additional universes.

Why UED Favors an Energy-Centered Approach Over Unnecessary Spatial Complexities

String Theory introduces higher dimensions, vibrating strings, and a multiverse landscape to explain fundamental forces. However, none of these are observable, and they add complexity rather than simplifying reality.

UED, in contrast, explains:

- Matter, forces, and interactions as energy transformations within the known four-dimensional universe.
- Gravity as an emergent energy equilibrium, rather than a force leaking into extra dimensions.
- Entanglement and quantum mechanics as structured energy processes, not mysterious connections through hyperspace.

1. Why Simplicity Matters in a Scientific Framework

- A successful theory should not introduce unnecessary entities (Occam's Razor).
- Extra dimensions are useful mathematically, but if reality can be explained without them, they may be an unnecessary abstraction.
- UED shows that structured energy flow accounts for all known quantum behaviors, making higher dimensions redundant.

2. The Testability Problem with String Theory

- String Theory requires energies far beyond our current experimental reach.
- Extra dimensions and vibrating strings have never been observed, making the theory speculative.
- UED, on the other hand, is built on observable energy dynamics, quantum correlations, and conservation laws.

❖ **UED Perspective:** A good theory explains the universe in the simplest terms possible. If energy dynamics alone can unify physics, there is no need to introduce hidden spatial structures.

UED vs. String Theory—A Shift in Perspective

String Theory attempts to unify forces by introducing extra dimensions and vibrating strings, but:

- Higher dimensions have never been observed, making the theory speculative.
- Gravity, quantum mechanics, and entanglement can all be explained without extra spatial structures.
- UED provides a simpler alternative—structured energy flow explains all fundamental interactions within the known four-dimensional space-time.

This leads to a profound realization:
If forces, matter, and reality itself emerge from structured energy flow, does this suggest that reality is a fundamentally interconnected system, rather than a collection of separate universes? Can we resolve the multiverse paradox by treating all "possible states" as structured energy configurations rather than independent worlds?
In the next section, we explore the paradox of parallel universes and the multiverse, revealing whether UED supports alternative realities or whether all of existence follows a single, interconnected energy framework.

6.6 The Paradox of Parallel Universes and the Multiverse in UED

Quantum mechanics has long suggested that reality may not be limited to a single universe. The Many-Worlds Interpretation (MWI), the concept of quantum superposition, and the probabilistic nature of wave functions all hint at the possibility of multiple coexisting universes. However, the nature of these universes—whether they are truly separate realities or just structured energy states within a single interconnected system—remains one of the biggest open questions in physics.

From a Universal Energy Dynamics (UED) perspective, reality is fundamentally structured by energy equilibrium, conservation, and symmetry. Rather than accepting an infinite branching multiverse where every possibility creates a separate reality, UED proposes that all possible states may exist as structured energy configurations within a single interconnected framework.
In this section, we will explore:

- Why quantum mechanics suggests the possibility of multiple universes.
- Can entanglement exist between universes, not just within one?
- If the universe is structured by energy equilibrium, do parallel universes exist as alternative energy states rather than independent realities?
- Does UED support a branching multiverse or a single, interconnected structure where all possibilities exist as energy configurations?

Why Does Quantum Mechanics Suggest the Possibility of Multiple Universes?

The idea of multiple universes largely arises from the probabilistic nature of quantum mechanics and wave function behavior.

1. The Many-Worlds Interpretation (MWI) and Quantum Superposition

- In classical physics, a system exists in one definite state at a time.
- In quantum mechanics, particles exist in superposition, meaning they hold multiple possible states simultaneously until measured.
- The Many-Worlds Interpretation (MWI) proposes that each quantum measurement causes a "split," creating multiple universes where every possibility is realized in a separate reality.

However, this raises fundamental questions:

- If every possibility creates a new universe, where does the energy come from to sustain infinite new realities?
- Does this mean reality is infinitely branching with no structured conservation of energy?
- How do these universes interact, or do they remain permanently isolated?

❖ **UED Perspective:** Rather than infinite separate universes, all possible states may exist as energy configurations within a structured, interconnected system, ensuring energy conservation.

Can Entanglement Exist Between Universes, Not Just Within One?

One of the most intriguing questions in quantum mechanics is whether entanglement could connect different universes. If entanglement preserves energy balance across space-time, could it also maintain balance between different quantum states that some interpret as separate universes?

1. Quantum Decoherence and Entangled Realities

- Quantum mechanics describes wave function collapse, where a superposed state resolves into a definite outcome.
- Some interpretations suggest that instead of collapsing, each possibility "branches" into a parallel reality.
- However, from a UED perspective, this may not mean a separate universe—it could simply be an alternative structured energy state within the same interconnected framework.

2. Is Entanglement a Universal Energy Link Rather Than a Localized Effect?

- If entanglement is a property of structured energy, it may not be confined within a single observable universe.
- Could multiple "versions" of reality be entangled as part of a larger, structured energy equilibrium system?
- If so, information and energy would not need to "jump" between universes—they would already be part of the same underlying system, just in different configurations.

❖ **UED Perspective:** Entanglement may not only connect particles across space-time but could also maintain structured energy balance across different "versions" of reality within a single interconnected system.

If the Universe Is Structured by Energy Equilibrium, Do Parallel Universes Exist as Alternative Energy States Rather Than Independent Realities?

Rather than treating parallel universes as physically separate worlds, UED proposes that they may be alternative configurations of the same underlying energy system.

1. Alternative Energy States vs. Branching Realities

- Instead of an infinite number of physically separate worlds, every quantum event may shift the system into a new energy configuration within the same underlying framework.
- This means different quantum outcomes do not create new universes but simply manifest as different structured energy states within the existing one.
- Just as a wave function exists in multiple states before measurement, reality itself may exist

in structured possibilities that interact rather than split.

2. The Conservation of Energy Across Quantum States

- A major issue with infinite universes is that it suggests an unlimited creation of energy with no conservation principles.
- UED maintains that all energy interactions must follow conservation laws, meaning that "parallel universes" (if they exist) must be part of a single conserved energy structure rather than independent, infinitely branching realities.

UED Perspective: Instead of infinitely branching realities, parallel universes (if they exist) are better understood as structured energy states that exist in an interconnected equilibrium rather than separate, isolated worlds.

Does UED Support a Branching Multiverse or a Single, Interconnected Structure?

String Theory and the Many-Worlds Interpretation suggest that every possible event creates a separate universe. However, this view raises problems related to energy conservation, causality, and testability. UED offers a different perspective:

1. The Universe as a Single, Structured Energy System

- Rather than branching universes, all quantum states exist as structured possibilities within an interconnected system.
- Instead of separate realities, all outcomes are energy fluctuations within the same universal equilibrium.
- This aligns with entanglement, conservation laws, and symmetry principles, eliminating the need for infinite physical universes.

2. UED's Energy-Based Alternative to the Multiverse

- The traditional multiverse model treats different universes as isolated realities with no direct connection.
- UED suggests that instead of separate worlds, all possible states are structured within a single energy network.
- Just as entanglement connects particles across space, all energy configurations may remain interconnected rather than existing independently.

❖ **UED Perspective:** The universe does not infinitely branch into independent realities—rather, all possible states exist within a single, structured energy equilibrium system.

UED and the Multiverse—A Shift in Understanding

The idea of multiple universes arises from quantum mechanics, but from a UED perspective, reality is not infinitely branching—it is a single, interconnected energy structure.

- Quantum mechanics does not require infinite universes—it can be explained as structured energy states.
- Entanglement may extend beyond individual particles, linking different versions of reality as energy equilibrium states.
- Parallel universes (if they exist) may not be separate worlds but alternative energy configurations within a single interconnected framework.
- UED does not support an infinite, branching multiverse but rather an organized, structured energy network where all possible states exist as part of a single conserved system.

This leads to a profound realization:
If reality follows a structured energy network, does this mean that intelligence, consciousness, and decision-making also exist as interconnected processes rather than isolated systems? Could thoughts, awareness, and perception operate under the same structured energy framework as quantum mechanics?
In the next section, we explore how UED resolves quantum mysteries and open questions, revealing why structured energy flow may explain not only physics but also intelligence, consciousness, and awareness.

6.7 Unraveling Other Quantum Mysteries: What UED Explains Without Paradoxes

Quantum mechanics is often described as paradoxical, filled with seemingly impossible behaviors that contradict classical physics. Phenomena like quantum tunneling, wave-function collapse, quantum gravity, and quantum computing challenge our conventional understanding of reality.
However, from the Universal Energy Dynamics (UED) perspective, these quantum mysteries are not paradoxes but structured energy interactions. Instead of treating quantum behavior as inexplicable or random, UED reveals that these effects are simply manifestations of energy equilibrium, conservation, and structured transformations at microscopic scales.
In this section, we will explore:

- Quantum tunneling: How particles cross energy barriers—structured energy transition rather than a violation of classical rules.
- Wave-function collapse: Energy reconfiguration, not a sudden "choice" of reality.
- Quantum gravity: Is gravity itself an entangled energy interaction rather than a fundamental force?
- Quantum computing: Does structured energy processing explain why quantum states perform complex calculations?

Quantum Tunneling: How Particles Cross Energy Barriers—Structured Energy Transition Rather Than a Violation of Classical Rules

In classical physics, a particle cannot pass through a barrier unless it has enough energy to overcome it. However, in quantum mechanics, particles can "tunnel" through barriers even when they seemingly lack the required energy. This phenomenon, known as quantum tunneling, is used in technologies such as scanning tunneling microscopes and nuclear fusion.

1. The Misconception of Tunneling as "Quantum Magic"

- Classical intuition suggests that a ball rolling up a hill must have enough energy to cross the peak.
- Quantum mechanics, however, allows particles to appear on the other side of a barrier without ever being observed climbing over it.
- This seems to violate classical energy conservation, leading to the misconception that tunneling allows particles to "teleport."

2. UED Explanation: Energy Redistribution, Not Magic

- Instead of violating classical laws, tunneling is simply an energy redistribution process within a structured system.
- The particle's wave function extends beyond the barrier, meaning that part of its energy configuration already exists on the other side.
- UED views tunneling as a natural consequence of structured energy flow, where energy is continuously redistributing rather than being confined to classical rules.

❖ **UED Perspective:** Quantum tunneling is not a violation of physics—it is an expression of structured energy transition, ensuring energy balance across a system rather than breaking classical rules.

Wave-Function Collapse: Energy Reconfiguration, Not a Sudden "Choice" of Reality

One of the most debated mysteries in quantum mechanics is wave-function collapse. When a quantum system is not observed, it exists in a superposition of multiple states. However, when measured, it "collapses" into a single outcome.

1. The Traditional View: Does Measurement "Decide" Reality?

- Quantum mechanics suggests that before observation, a particle is in a probability wave of multiple possible states.
- The act of measurement causes the wave function to "collapse" into a single, definite outcome.
- This has led to interpretations like the Copenhagen interpretation, Many-Worlds interpretation, and observer-dependent reality models.

2. UED Explanation: Measurement as Energy Reconfiguration, Not an Arbitrary Choice

- Instead of seeing wave-function collapse as a forced decision between multiple realities, UED treats it as a structured energy reconfiguration.
- The system was already in an energy-balanced state before measurement, and the act of observation simply finalizes that structured energy distribution.
- Rather than causing collapse, measurement aligns energy equilibrium in a defined way, making the final state observable.

❖ **UED Perspective:** Wave-function collapse is not a sudden "choice" of reality but an energy stabilization process, ensuring that a quantum system reaches a structured energy state upon interaction.

Quantum Gravity: Is Gravity Itself an Entangled Energy Interaction Rather Than a Fundamental Force?

One of the biggest open problems in physics is how to reconcile gravity with quantum mechanics. Unlike the other three fundamental forces (electromagnetism, strong nuclear, weak nuclear), gravity does not fit into the Standard Model of quantum physics.

1. The Problem with Treating Gravity as a Fundamental Force

- General relativity describes gravity as a curvature of space-time caused by mass.
- However, gravity does not behave like other forces—it does not have a known quantum

particle (graviton), and it does not exhibit the same quantum behaviors.

- If gravity were a fundamental force, it should have a quantum interaction model like electromagnetism, but no such model has been confirmed.

2. UED Explanation: Gravity as an Emergent Energy Balance Mechanism

- Instead of treating gravity as a force, UED proposes that gravity is an emergent effect of structured energy equilibrium.
- Gravity may not require force carriers (gravitons)—it may be an entangled energy system that redistributes mass-energy across space-time.
- This aligns with our earlier discussion on gravity as an energy distribution process rather than just space-time curvature.

❖ **UED Perspective:** Gravity may not be a fundamental force—it may be a structured energy interaction that maintains equilibrium, similar to entanglement, rather than requiring a force carrier.

Quantum Computing: Does Structured Energy Processing Explain Why Quantum States Perform Complex Calculations?

Quantum computing operates using qubits, which exist in multiple states simultaneously (superposition) and interact in correlated ways (entanglement). This allows quantum computers to solve problems exponentially faster than classical computers.

1. The Mystery of Quantum Computing's Exponential Power

- Classical computers process information using binary states (0s and 1s).
- Quantum computers use superposition and entanglement to process multiple possibilities at once, exponentially increasing computational efficiency.
- This raises the question: Why does quantum computation behave so differently from classical computation?

2. UED Explanation: Quantum Computation as Structured Energy Flow

- Instead of viewing quantum computing as an isolated quantum effect, UED suggests that it harnesses structured energy interactions to optimize problem-solving.
- Entanglement and superposition allow information to be processed in a balanced energy system, ensuring that multiple possibilities exist simultaneously before reaching a final state.

- This suggests that intelligence itself may function similarly, using structured energy to process and optimize decisions.

❖ **UED Perspective:** Quantum computing is not just a technical breakthrough—it reveals how structured energy flow optimizes problem-solving, possibly hinting at the way intelligence and thought operate.

UED Resolves Quantum Mysteries without Paradoxes

From the UED perspective, quantum mechanics is not paradoxical—it is an expression of structured energy interactions that maintain equilibrium across all scales.

- Quantum tunneling is a structured energy transition, not a violation of classical rules.
- Wave-function collapse is an energy reconfiguration process, not a sudden "choice" of reality.
- Gravity may not be a fundamental force but an entangled energy balancing system.
- Quantum computing functions as a structured energy processor, optimizing solutions through entanglement and coherence.

This leads to a profound realization:
If quantum mechanics follows structured energy equilibrium, does this principle extend to intelligence, decision-making, and perception? Could thought itself be a quantum-like structured energy process?
In the next section, we explore how UED connects quantum behavior with intelligence and awareness, revealing how structured energy may govern the emergence of thought, perception, and decision-making.

6.8 Conclusion: Entanglement as the Energy Web of Reality

Quantum mechanics has long challenged our classical understanding of reality, presenting concepts like entanglement, superposition, wave-function collapse, and quantum uncertainty—all of which defy traditional notions of space, time, and causality. However, when analyzed through the lens of Universal Energy Dynamics (UED), these so-called paradoxes are not contradictions, but structured energy processes that maintain equilibrium across all scales.
Entanglement is often considered one of the most mysterious quantum phenomena, but UED reveals that it is not a paradox—it is a structured energy network that ensures universal connectivity and energy balance. Instead of viewing entanglement as an anomaly that violates locality, we now understand that it is simply the manifestation of deeper energy symmetry and conservation principles.
This chapter has demonstrated that:

- Quantum entanglement is not a paradox—it is a structured energy network.
- Information, energy, and consciousness may all be interconnected through entanglement.
- UED expands quantum mechanics by showing that entanglement is an expression of structured energy equilibrium.
- Multiverse, string theory, and other quantum mysteries should be reevaluated based on energy balance rather than unnecessary complexity.

Quantum Entanglement as an Organized Energy Network

Rather than assuming entanglement is a strange exception to classical physics, UED suggests that it is a fundamental aspect of how energy systems maintain structure across time and space.

- Entangled particles do not "communicate" faster than light—they remain part of a structured energy configuration that ensures zero-net equilibrium.
- Energy conservation principles require that once two systems interact, their energy remains correlated regardless of distance.
- The universe is not a collection of disconnected particles but a network of structured energy states that continuously interact and rebalance.

Thus, entanglement is not a violation of physics but proof that the universe operates as a connected energy system.

Information, Energy, and Consciousness as Interconnected Structures

The deeper implication of entanglement is that information is not separate from energy—it is structured within energy itself.

- Quantum information is stored in entangled systems, suggesting that energy inherently encodes structure and meaning.
- If consciousness and thought patterns follow structured energy flow, could they also be part of a larger entangled system?
- The brain functions using electromagnetic and quantum-level interactions, indicating that awareness could be an expression of structured energy dynamics rather than a purely biological function.

This means that entanglement may not only exist at the level of particles but could extend to information networks, intelligence, and even the structure of awareness itself.

Reevaluating Quantum Mysteries with Energy Balance Instead of Complexity

Many existing quantum theories attempt to explain reality by introducing new dimensions, infinite branching universes, or hidden variables. However, these models often add unnecessary complexity without resolving fundamental paradoxes.

1. The Multiverse May Be an Energy System, Not an Infinite Branching Reality

- Instead of every possibility spawning a new universe, UED suggests that all possible states exist as structured energy configurations within the same interconnected system.
- Energy conservation implies that realities are not endlessly splitting, but existing in balanced energy distributions.

2. String Theory May Be Overcomplicating Reality with Extra Dimensions

- Rather than requiring 10 or 11 dimensions to unify physics, UED proposes that structured energy flow accounts for quantum behavior without hidden spatial dimensions.
- If energy inherently follows balance and symmetry, why introduce unseen dimensions to explain what can be described using known energy principles?

3. Quantum Gravity as an Entangled Energy Interaction Rather Than a Force

- Gravity has resisted quantization, leading to speculation about gravitons and extra-dimensional forces.
- UED suggests that gravity is not a fundamental force but an emergent property of structured energy seeking equilibrium across mass-energy distributions.
- If gravity is inherently connected to entanglement, it may not require force carriers but function as a universal balancing mechanism.

Thus, rather than treating quantum mechanics as an isolated domain with inexplicable effects, UED integrates quantum behavior into a larger structured framework of energy equilibrium, reducing unnecessary complexity.

The Future of UED: Moving Beyond Quantum Physics into Intelligence and Consciousness

If entanglement reveals that reality is an interconnected energy system, then the next major question is:

Does intelligence itself emerge from structured energy flow? Is thought an entangled energy state interacting with the universe?

Quantum mechanics has already suggested that observation affects reality, that measurement finalizes energy configurations, and that information is embedded in energy. If these principles apply to fundamental particles, could they also apply to intelligence, perception, and decision-making?
Having now established UED's role in physics and quantum mechanics, our next step is to explore:

- The relationship between structured energy and biological evolution.
- How intelligence and thought emerge as structured energy interactions.
- Whether consciousness is an entangled system within universal energy balance.

With this foundation, we are now prepared to move beyond quantum mechanics and into the realm of life, intelligence, and awareness—where energy does not simply organize physical matter, but also shapes perception, decision-making, and the emergence of self-awareness.

Chapter 7: The Structure of the Universe – The Energy Medium

As we have explored throughout this book, Universal Energy Dynamics (UED) provides a comprehensive framework that unifies quantum mechanics, classical physics, and energy equilibrium principles. Now, we extend this perspective to the largest scale possible—the structure of the universe itself.

One of the biggest misconceptions in modern physics is that space is "empty." However, experimental evidence—ranging from the effects of dark matter and dark energy to vacuum fluctuations and quantum fields—suggests that the universe is filled with a structured energy medium. This underlying Universal Energy Medium (UEM) governs the evolution, expansion, and stability of the cosmos.

Through UED, we will answer some of the biggest questions in cosmology:

- Does space have an intrinsic energy structure?
- What is the real nature of dark matter and dark energy?
- Is cosmic expansion a function of energy redistribution rather than space-time stretching?
- Are black holes energy recycling zones rather than singularities of infinite density?
- Does UED provide a new perspective on the Big Bang, cosmic origins, and the ultimate fate of the universe?

This chapter will show that rather than being a collection of disconnected theories, the universe is a single interconnected energy system governed by structured transformations, balance, and conservation.

7.1 The Universe Is Not Empty – The Universal Energy Medium (UEM)

For centuries, classical physics has treated space as a vast, empty void—a stage upon which celestial bodies move, governed by gravity and fundamental forces. This assumption was rooted in Newtonian mechanics, which viewed space as an absolute emptiness, containing matter but possessing no intrinsic properties of its own. Even Einstein's general relativity, which describes space-time as a flexible fabric, did not attribute any inherent energy structure to it beyond curvature.

However, modern physics—particularly quantum field theory (QFT) and cosmology—has shown that space is not empty at all. Instead, it is filled with vacuum energy, quantum fluctuations, and possibly an underlying field structure that governs all cosmic interactions.

From the perspective of Universal Energy Dynamics (UED), space itself is not "nothingness"—it is an active, structured energy medium that maintains universal equilibrium. This Universal Energy Medium (UEM) is not just a passive backdrop but the foundation of all forces, interactions, and cosmic balance.

In this section, we will explore:

- Why classical physics assumed space was a vacuum.
- How quantum field theory already suggests that space is filled with structured energy interactions.
- UED's perspective: Space is an energy medium, not emptiness, ensuring universal equilibrium.
- Could UEM be the foundation of all cosmic forces and interactions?

Why Classical Physics Assumed Space Was a Vacuum

For much of scientific history, space was considered "nothing"—a void through which planets, stars, and galaxies move.

1. Newton's View: Absolute Empty Space

- Newtonian mechanics treated space as a passive stage where objects exert gravitational forces on one another.
- There was no need for space itself to have any structure—only the masses within it were thought to influence motion.
- This led to the belief that space was "empty" and devoid of energy.

2. Einstein's Space-Time: Still a Geometric Concept, Not an Energy Medium

- Einstein's general relativity introduced the idea that space-time is curved by mass-energy, but space itself was still viewed as a geometric construct rather than an active energy field.
- While general relativity describes space-time as dynamic, responding to mass-energy, it does not propose that space itself possesses intrinsic energy.

3. Classical Vacuum: The Misconception of "Empty Space"

- Classical physics defined a vacuum as an absence of matter.

- The assumption was that if no particles existed in a region of space, then no physical properties could be assigned to that space.
- This led to the belief that space is "nothingness"—a complete void rather than an energy-rich medium.

❖ **UED Perspective:** Classical physics missed a fundamental truth—space is not empty but an energy medium, influencing all interactions in the universe.

How Quantum Field Theory Already Suggests That Space Is Filled with Structured Energy Interactions

Modern physics has overturned the classical vacuum model, showing that space is not empty but filled with quantum fluctuations, vacuum energy, and possibly hidden energy structures.

1. Quantum Fluctuations: Energy Exists Even in Empty Space

- Quantum mechanics predicts that even in the most "empty" vacuum, particles and antiparticles constantly pop in and out of existence.
- These fluctuations are real and have measurable effects, such as:
    - **The Casimir Effect:** Demonstrates that quantum fluctuations exert force even in a vacuum.
    - **Hawking Radiation:** Suggests that black holes evaporate due to virtual particle interactions in the vacuum.
- This proves that space itself has energy, even when no classical particles are present.

2. Vacuum Energy and the Cosmological Constant Problem

- The vacuum energy density predicted by quantum mechanics is enormous, yet cosmological observations suggest a much smaller energy density (linked to dark energy).
- This discrepancy suggests that space contains an underlying energy field that we have not fully understood.

3. The Higgs Field: Proof That Space Has an Active Energy Structure

- The Higgs field permeates all of space, giving particles mass through energy interactions.
- This suggests that other fundamental fields could also fill space, interacting with energy and matter in ways we have yet to explore.

❖ **UED Perspective:** Quantum mechanics has already confirmed that space is not empty—it contains structured energy, which aligns with the UEM model of UED.

UED's Perspective: Space Is an Energy Medium, Not Emptiness, Ensuring Universal Equilibrium

Rather than treating space as an empty void, UED proposes that space itself is a structured energy medium that ensures the balance and stability of all cosmic interactions. This Universal Energy Medium (UEM) acts as the foundational field that organizes energy transformations, maintains symmetry, and facilitates connectivity at all scales.

1. UEM as the Foundation of All Forces

- Instead of viewing gravity, electromagnetism, and nuclear forces as independent interactions, UED suggests they are all expressions of energy equilibrium within the UEM.
- Forces arise not from empty space, but from structured energy redistributions within this medium.

2. UEM as the Explanation for Dark Energy and Dark Matter

- Dark energy may not be an unknown force—it could be a structured energy field within UEM driving cosmic expansion.
- Dark matter may not be exotic particles—it could be the result of energy concentrations within the UEM that do not interact electromagnetically.

3. UEM as the Medium for Entanglement and Quantum Connectivity

- If entanglement is a universal energy connection, it requires a structured medium to maintain symmetry across vast distances.
- The UEM provides the energy network that allows entanglement, coherence, and quantum states to remain stable.

❖ **UED Perspective:** Space itself is an active, structured energy system, ensuring universal equilibrium rather than being an empty void.

Could UEM Be the Foundation of All Cosmic Forces and Interactions?

If space is not empty but a structured energy field, then it must play a fundamental role in all cosmic interactions.

1. Could UEM Be the Unifying Medium of Physics?

- The Standard Model describes fundamental forces without explaining why they emerge from space.
- If space is an energy medium, then all forces could be interpreted as energy redistributions within the UEM.

2. Could UEM Be the Missing Link Between Gravity and Quantum Mechanics?

- One of the biggest problems in physics is that gravity does not fit within quantum mechanics.
- If UEM is a universal energy network, then gravity may not be a separate force but an energy redistribution effect within this medium.

3. Could UEM Explain Cosmic Expansion Without Dark Energy?

- If the universe is expanding, what is causing space to stretch?
- Rather than invoking dark energy as an unknown force, UED suggests that expansion is an energy redistribution effect within the UEM, ensuring large-scale equilibrium.

❖ **UED Perspective:** Rather than searching for new forces, new dimensions, or missing mass, modern physics should focus on understanding the UEM as the foundation of all interactions.

Space as an Energy Medium, Not a Void

From a UED perspective, space is not empty—it is an interconnected energy network that governs all forces and cosmic interactions.

- Classical physics wrongly assumed space was a void, but quantum mechanics has shown that energy exists even in "empty" space.
- The Universal Energy Medium (UEM) provides a structured framework that maintains cosmic balance and energy flow.
- UEM could be the foundation of dark matter, dark energy, and quantum entanglement, offering a unified view of cosmic structure.
- Understanding UEM may be the key to unifying gravity, quantum mechanics, and cosmology without the need for new particles or extra dimensions.

This leads to a profound realization:
If space itself is an energy medium, does this mean that dark matter, dark energy, black holes, and cosmic expansion are simply structured interactions within this network rather than independent mysteries?
In the next section, we explore how UED provides a new perspective on dark matter and dark energy, showing that they are not exotic entities but manifestations of energy equilibrium within the UEM.

7.2 Microbial Survival in Space as Evidence of an Energy Medium

For decades, space has been considered a harsh, life-inhibiting environment characterized by extreme radiation, vacuum, and temperature fluctuations. However, multiple experiments have demonstrated that certain microorganisms, such as bacteria and extremophiles, can survive prolonged exposure to space conditions. These findings challenge the traditional view of space as an empty void and suggest that a structured energy medium exists, supporting biological stability in unexpected ways.
Under Universal Energy Dynamics (UED), microbial survival in space is not an anomaly but a confirmation that space is not empty—it contains structured energy interactions that stabilize biological matter.

The Experimentally Proven Survival of Microorganisms in Space

Several space missions have tested whether life can withstand the extreme conditions of outer space, and the results consistently show that certain life forms persist even without protection from an atmosphere.

- **Tardigrades (Water Bears)** – Survived direct exposure to space on ESA's FOTON-M3 mission.
- **Deinococcus radiodurans (Radiation-Resistant Bacteria)** – Survived for years in low-Earth orbit experiments.
- **Lichen and Fungal Spores** – Survived vacuum exposure in ESA's EXPOSE-E experiment on the ISS.

Key UED Insight: If space were truly "empty," life should immediately decay, lose molecular integrity, and be destroyed by radiation. The fact that microorganisms can persist in vacuum conditions suggests the presence of an underlying energy structure that supports biological stability.

How Do Microorganisms Survive in Vacuum?

Microbes do not survive space simply by being "tough." Their resilience suggests that they interact with an invisible energy medium that stabilizes molecular structures.

A. Quantum-Level Energy Support Mechanisms

- Some extremophiles enter cryptobiosis, a state where all metabolic functions halt, yet their molecular integrity remains intact.
- This suggests that they might be using quantum-level energy fluctuations in vacuum to prevent total molecular collapse.
- Quantum field theory already shows that vacuum is not empty—it contains zero-point energy that may interact with biological molecules, preventing their destruction.

B. Structured Energy Fields as Biological Shields

- Space is filled with electromagnetic and quantum vacuum fluctuations.
- Certain microorganisms may use these invisible energy interactions as a stabilization mechanism, shielding them from the destructive effects of space.
- Just as plants use photosynthesis to structure light energy, microbes may utilize energy field structuring for survival.

Key UED Insight: This aligns with UED's principle that energy equilibrium maintains order in all systems, even in extreme environments like space. If life can survive in space through energy interactions, then space itself must contain a structured energy medium that interacts with biological systems.

Implications for the Universal Energy Medium (UEM)

The ability of life to persist in space has profound implications for the nature of space itself.

- If space were truly empty, it would be completely inhospitable to life at any scale.
- The fact that biological structures can remain stable in vacuum conditions suggests that space is not empty but filled with an energy medium that interacts with matter.

Key UED Insight: Microbial survival in space serves as direct experimental evidence that space contains structured energy interactions. This supports the idea that the Universal Energy Medium (UEM) governs cosmic energy balance, matter stability, and even the potential for life beyond Earth.

Could Space Itself Be a Living System?

If life interacts with structured energy fields in space, does this mean that space itself has biological properties?

- If energy structures support biological survival, could they also enable biological processes to emerge beyond Earth?
- Just as cells operate within structured biological fields, could the universe itself function as

an interconnected energy-organism?

- This supports theories like panspermia—the idea that life can naturally spread through space due to structured energy interactions.

Key UED Insight: If space itself sustains biological stability, then life is not an anomaly but an inevitable result of structured energy flow. This could redefine how we search for extraterrestrial life—not just looking for planets, but for structured energy fields that support life.

Final Conclusion: Microbial Survival Proves Space Is Not Empty

- The survival of microbes in space is not an anomaly—it is evidence that an invisible energy medium exists, stabilizing biological matter.
- Quantum energy fluctuations and structured electromagnetic interactions might play a role in sustaining molecular integrity even in extreme environments.
- The Universal Energy Medium (UEM) is not just a theoretical concept—it is experimentally supported by the ability of life to persist in vacuum conditions.

7.3 The Womb Analogy: Matter Formation as a Structured Process

One of the greatest mysteries in physics is how matter emerged from a seemingly homogeneous energy state. Conventional cosmology often describes the early universe as a chaotic, high-energy explosion (the Big Bang), where particles appeared spontaneously. However, this abrupt origin model fails to explain how structured systems—such as galaxies, atoms, and even life—emerged from an initial singularity.

Universal Energy Dynamics (UED) provides a more refined perspective: matter did not form instantaneously, but through a structured energy transformation process. To illustrate this, we introduce a biological parallel—the womb analogy, which conceptualizes the formation of matter not as an explosion, but as a gradual process of structured emergence.

The Universe as a Womb: The Gradual Formation of Matter

The development of matter in the early universe can be compared to the way a fetus develops inside the womb. Just as an embryo does not suddenly appear as a fully formed human but gradually takes shape within a controlled, structured environment, matter in the early universe emerged within a stable energy field, rather than appearing out of nothing.

In the Womb: A fertilized egg begins as a simple cell and gradually differentiates into complex tissues and organs, following a structured process governed by genetic instructions and biochemical energy regulation.

In the Universe: Matter did not emerge chaotically—it formed within a structured energy equilibrium, where quantum fields gradually differentiated into particles, forces, and large-scale cosmic structures.

Just as a womb provides a stable, nutrient-rich environment for controlled development, the early universe functioned as an energy medium, ensuring that matter formed in a structured, gradual, and optimized manner.

B. Why the Big Bang Model Fails to Explain Matter Formation

The standard Big Bang Theory suggests that the universe began as a singularity and expanded explosively, creating matter through rapid cooling. However, this raises several contradictions:

- If the **Big Bang** was purely an explosion, why did matter structure itself so precisely rather than dispersing randomly?
- Why do we observe galaxies, atoms, and life as organized systems rather than chaotic debris?
- What provided the initial framework for stable matter formation?

In contrast, UED suggests that matter emerged not from an uncontrolled explosion but from a stable energy field, much like how a fetus grows within the structured energy of a womb.

- **The First Disturbance** triggered structured energy interactions, much like fertilization triggers embryonic development.
- Quantum fluctuations were not chaotic—they followed structured energy patterns, leading to stable matter formation.
- The emergence of fundamental forces acted as a self-regulating system, ensuring that matter could develop in an organized manner.

Structured Energy Fields as the Universal Womb

If the universe functioned like a womb, then it must have contained a pre-existing energy medium capable of structuring matter in a stable manner. This aligns with UED's concept of the Universal Energy Medium (UEM), which serves as a structured field where energy differentiation gives birth to physical reality.

How the Womb Analogy Explains Universal Energy Medium (UEM):

- Just as the amniotic sac provides a protective environment for fetal growth, the UEM provided a balanced energy equilibrium for the emergence of matter.
- Just as DNA guides fetal development, structured energy laws governed the formation of forces, particles, and galaxies.
- Just as the mother's body ensures energy efficiency in fetal growth, the universal energy medium regulated the balance between matter and energy, ensuring the stability of cosmic structures.

Thus, matter formation was not accidental—it was an organized, structured process, governed by energy equilibrium.

The Role of Black Holes as the Universe's Placental System

In biological development, the placenta plays a critical role in nutrient exchange, filtering waste, and maintaining equilibrium between the fetus and the external environment. Similarly, in the universe, black holes act as cosmic regulators, redistributing energy and matter to maintain balance.
How Black Holes Function as a Placental System:

- **Energy Inflow:** Just as a placenta transfers nutrients from the mother to the fetus, black holes absorb energy and matter, preventing uncontrolled energy buildup.
- **Energy Redistribution:** The placenta regulates oxygen and nutrient flow, while black holes release structured energy through relativistic jets, ensuring cosmic balance.
- **Matter Transformation:** The placenta helps convert nutrients into usable forms for fetal development, just as black holes transform absorbed matter, feeding the cosmic environment with new elements.

Thus, black holes are not destructive endpoints—they are dynamic regulators that maintain the equilibrium of the structured universe.

The Ongoing Process: The Universe is Still in Its Womb Phase

A crucial implication of this analogy is that matter formation is not a past event—it is an ongoing structured process. Just as the fetal development stage eventually leads to birth, the universe is still in a continuous cycle of structured energy transformations.

- New galaxies, stars, and planetary systems continue to form, following the same structured energy principles that shaped the early universe.
- Black holes and cosmic structures continue to recycle matter and energy, ensuring perpetual

cosmic balance.
- The presence of dark matter-like effects suggests that structured energy fields are still influencing matter formation.

Thus, the universe is not a completed system—it is still undergoing structured transformations, much like a fetus still developing in the womb.

The Womb Analogy as a Unifying Model for Matter Formation

- Matter did not emerge through a chaotic explosion—it followed a structured energy transformation process, much like biological development in a womb.
- The Universal Energy Medium (UEM) acted as a structured energy field that enabled the emergence of particles, forces, and galaxies in an organized manner.
- Black holes function as cosmic regulators, maintaining energy equilibrium and ensuring the continuous cycle of matter transformation.
- The universe is still evolving—matter is still forming, stars are still being born, and energy structuring continues in an endless cycle of cosmic evolution.

7.4 How UED Explains Dark Matter and Dark Energy – Hidden Energy Fields Balancing the Cosmos

For decades, cosmologists have struggled to explain two of the biggest mysteries in astrophysics: dark matter and dark energy. These unseen components are thought to make up 95% of the universe's total mass-energy content, yet they remain undetectable through direct observation.

- Dark matter is inferred through its gravitational effects on galaxies, but no one has ever directly observed a dark matter particle.
- Dark energy is hypothesized to drive the accelerating expansion of the universe, but its exact nature remains unknown.

Mainstream physics treats these as two separate phenomena requiring new particles, unknown forces, or exotic physics beyond the Standard Model. However, Universal Energy Dynamics (UED) provides a different perspective: dark matter and dark energy are not exotic—they are manifestations of structured energy balance within the Universal Energy Medium (UEM).
This section explores:

- The mystery of dark matter: Does it actually exist, or is it an effect of structured energy

equilibrium?
- UED's interpretation of dark matter as energy networks maintaining cosmic balance rather than invisible particles.
- The enigma of dark energy: Is the universe accelerating due to an unknown force, or is this simply an effect of large-scale energy redistribution?
- How structured energy flow eliminates the need for "missing mass" explanations in galaxies.

The Mystery of Dark Matter: Does It Actually Exist, or Is It an Effect of Structured Energy Equilibrium?

1. The Traditional View: Dark Matter as an Undetectable Substance

- Astronomers first proposed dark matter to explain why galaxies rotate faster than expected based on visible mass.
- According to Newtonian gravity and general relativity, stars at the edges of galaxies should orbit more slowly than those near the center—but observations show they move at nearly constant speeds.
- The conventional explanation is that galaxies contain huge amounts of unseen mass (dark matter) that provides the extra gravitational pull.

However, no direct detection of dark matter particles has ever been confirmed. All evidence is gravitational, meaning we might be misinterpreting its cause.

2. UED's Perspective: Dark Matter as an Energy Equilibrium Effect

- Instead of requiring an unknown particle, UED proposes that dark matter effects arise from structured energy networks maintaining cosmic balance.
- The Universal Energy Medium (UEM) distributes energy in ways that are not directly observable electromagnetically but still influence gravity.
- This means that the "missing mass" is not a separate form of matter but an emergent effect of structured energy maintaining equilibrium within galaxies.

❖ **UED Perspective:** Dark matter does not need to be a new type of particle—it can be explained as an energy balance mechanism within the UEM, ensuring stability in galaxy rotation.

UED's Interpretation of Dark Matter as Energy Networks Maintaining Cosmic Balance

If dark matter is not a particle but an effect of structured energy equilibrium, then its properties should align with this interpretation:

1. Why Dark Matter Does Not Interact Electromagnetically

- Traditional physics suggests that dark matter does not emit, absorb, or reflect light, making it invisible.
- UED suggests that what we call "dark matter" is actually structured energy fields in the UEM that interact gravitationally but not electromagnetically.

2. Why Dark Matter Forms Halos Around Galaxies

- Observations show that dark matter does not cluster like normal matter—it forms diffuse halos around galaxies.
- This aligns with UED's view that energy equilibrium structures extend beyond visible matter, maintaining gravitational balance across large scales.

3. Why No Dark Matter Particles Have Been Found

- If dark matter were a particle, experiments like the Large Hadron Collider (LHC) or deep-space surveys should have detected it by now.
- The failure to find dark matter particles suggests that it is not a separate type of matter but a structured energy effect.

❖ **UED Perspective:** Dark matter is not an exotic substance—it is a structured energy distribution within the UEM, stabilizing galaxies without requiring new fundamental particles.

The Enigma of Dark Energy: Is the Universe Accelerating Due to an Unknown Force, or Is This Simply an Effect of Large-Scale Energy Redistribution?

The discovery that the universe's expansion is accelerating led physicists to hypothesize dark energy—an unseen force driving space apart.

1. The Cosmological Constant and the Expanding Universe

- Einstein originally introduced the cosmological constant (Λ) to counteract gravity and maintain a static universe.
- Later, observations of supernovae showed that the universe is not just expanding—it is accelerating.

- This led to the theory that dark energy is some unknown force pushing galaxies apart.

However, no physical explanation for dark energy has been confirmed—it is simply an observational phenomenon that requires a theoretical cause.

2. UED's Perspective: Expansion as a Natural Energy Redistribution Process

- Instead of assuming an unknown force, UED suggests that cosmic expansion is simply a large-scale energy equilibrium process.
- The Universal Energy Medium (UEM) naturally redistributes energy over cosmic distances, ensuring balance across the universe.
- Rather than dark energy "pushing" space apart, expansion is a structured energy flow process that emerges from energy conservation laws.

❖ **UED Perspective:** The universe is not expanding due to a mysterious force—it is evolving as a self-regulating energy system within the UEM.

How Structured Energy Flow Eliminates the Need for "Missing Mass" Explanations in Galaxies

The biggest motivation for dark matter and dark energy theories is that galaxies do not behave as expected based on visible matter alone. However, if space is an energy medium, then:

- The "missing mass" in galaxies is not matter—it is the structured energy balance within the UEM.
- The accelerated expansion of the universe is not due to a force but a natural energy redistribution process.
- Cosmic structures follow equilibrium principles, ensuring energy remains balanced rather than accumulating in one location.

1. Dark Matter and Dark Energy as Different Manifestations of UEM

- Dark matter effects (extra gravity in galaxies) and dark energy effects (accelerated expansion) are opposite sides of the same energy balance mechanism.
- Both arise because space is not empty—it is an energy medium maintaining equilibrium.
- There is no need to propose separate particles or forces—all effects emerge from the structured dynamics of energy within UEM.

❖ **UED Perspective:** Instead of treating dark matter and dark energy as independent mysteries, UED sees them as different aspects of structured energy interactions, eliminating the need for missing mass explanations.

Dark Matter and Dark Energy as Structured Energy Balance Effects

From a UED perspective, dark matter and dark energy are not exotic entities requiring new physics. Instead, they are emergent effects of structured energy balance within the Universal Energy Medium (UEM).

- Dark matter does not need to be a particle—it is an energy network maintaining cosmic stability.
- Dark energy does not need to be a force—it is a large-scale energy redistribution process within the UEM.
- Galaxies behave as they do because energy equilibrium structures maintain balance beyond visible matter.
- The universe is evolving as a self-regulating energy system, not being "pushed" apart by an unknown force.

This leads to a profound realization:
If dark matter and dark energy are just different manifestations of structured energy flow, could black holes be another form of cosmic energy regulation rather than singularities of infinite density?
In the next section, we explore how UED redefines black holes—not as singularities but as energy storage and recycling zones that play a critical role in the universe's structured energy balance.

7.5 Cosmic Expansion – Energy Redistribution, Not Just Space-Time Stretching

One of the greatest challenges in modern cosmology is explaining why the universe is expanding and why this expansion appears to be accelerating. Traditional physics describes this process as space-time "stretching" due to an unknown force (dark energy), but it does not provide a clear mechanism for how or why this happens.
From a Universal Energy Dynamics (UED) perspective, the expansion of the universe is not caused by "stretching space" in a geometric sense, but rather by a structured energy redistribution process occurring within the Universal Energy Medium (UEM). The universe does not expand due to an external force—it evolves as a self-regulating energy system, maintaining equilibrium on a cosmic scale.
This section explores:

- Why traditional physics treats the universe as "stretching" without a clear mechanism.
- UED's explanation: The universe expands as a structured energy redistribution process, not because of space "stretching."
- How energy equilibrium prevents infinite expansion or contraction, ensuring a cyclic, self-sustaining structure.
- Could the universe breathe in structured cycles, rather than expand infinitely or collapse?

Why Traditional Physics Treats the Universe as "Stretching" Without a Clear Mechanism

Observations of distant galaxies show that light from them is redshifted, meaning they are moving away from us. This led to the idea that space itself is expanding, rather than galaxies simply moving apart.

1. The Big Bang Model and Space-Time Expansion

- The standard Big Bang model suggests that the universe began as a singularity and has been expanding ever since.
- General relativity describes space-time as a dynamic fabric, meaning that as the universe expands, distances between galaxies increase, even without motion through space.
- However, this does not explain why expansion is happening—only that it is observed.

2. The Introduction of Dark Energy to Explain Accelerated Expansion

- In the late 1990s, astronomers discovered that the universe's expansion is not slowing down but accelerating.
- To explain this, physicists proposed dark energy, an unknown force supposedly driving the expansion.
- However, dark energy remains an abstract mathematical term with no physical explanation.

❖ **UED Perspective:** Instead of assuming space "stretches" due to an unknown force, UED proposes that expansion is simply a result of structured energy redistribution across the Universal Energy Medium (UEM).

UED's Explanation: The Universe Expands as a Structured Energy Redistribution Process

Instead of treating expansion as a geometric property of space-time, UED suggests that expansion is driven by energy equilibrium processes within the UEM.

1. The Universe as an Energy Balancing System

- The UEM is not empty space but a structured energy medium, ensuring cosmic equilibrium.
- As energy redistributes across large scales, the universe naturally expands to maintain balance, rather than being "stretched" by an unknown force.
- Expansion is not random—it follows structured energy dynamics that regulate density variations.

2. Expansion as a Pressure-Driven Effect of Energy Balance

- Instead of an external force, cosmic expansion is the result of energy pressure differentials within the UEM.
- Just as heat expands a gas, energy shifts within the UEM naturally drive expansion at different rates depending on the energy density.
- This removes the need for dark energy as an independent force—it is simply an emergent effect of structured energy dynamics.

3. Why the Expansion Accelerates Without Requiring Extra Energy

- Traditional models assume an additional energy source is needed to cause acceleration.
- UED suggests that acceleration is not caused by "new" energy but by the way existing energy redistributes across large cosmic scales.
- The rate of expansion changes because energy flow within the UEM follows structured equilibrium adjustments over time.

❖ **UED Perspective:** Expansion is not a mysterious force stretching space—it is an intrinsic effect of energy balance within the UEM, following structured redistribution principles.

How Energy Equilibrium Prevents Infinite Expansion or Contraction, Ensuring a Cyclic, Self-Sustaining Structure

One of the biggest concerns in modern cosmology is the fate of the universe. If expansion continues indefinitely, the universe could become cold and lifeless (heat death). Alternatively, if gravity dominates, the universe could collapse in a Big Crunch. UED provides a third alternative—a self-sustaining cyclic model where expansion and contraction are part of a natural energy balancing process.

1. Why the Universe Will Not Expand Forever

- If expansion were purely geometric, it could continue indefinitely.
- However, UED suggests that the UEM acts as a self-regulating system, ensuring that expansion is balanced over time.
- Rather than infinite expansion, cosmic energy shifts over long timescales will naturally lead to a slowdown and eventual reconfiguration.

2. Why the Universe Will Not Collapse Into a Big Crunch

- Traditional models assume that if gravity overcomes expansion, the universe could contract into another singularity.
- UED suggests that such a collapse is unlikely because energy is always structured in a way that prevents absolute density imbalances.
- The UEM ensures that energy redistributes before the universe reaches an extreme state of collapse.

3. Cosmic Expansion as a Self-Sustaining Energy Cycle

- Instead of a universe that expands infinitely or collapses completely, UED suggests that the universe undergoes structured phases of expansion and stabilization.
- Energy equilibrium ensures that large-scale cosmic structures adjust dynamically rather than following a linear path to heat death or collapse.
- This aligns with recent observations that some cosmic structures seem to be maintaining a constant density rather than expanding infinitely.

❖ **UED Perspective:** The universe is neither "stretching infinitely" nor "collapsing"—it is evolving as a self-sustaining energy system that undergoes structured cycles of expansion and redistribution.

Could the Universe Breathe in Structured Cycles, Rather Than Expand Infinitely or Collapse?

Instead of treating expansion as a one-way process, UED proposes a cyclic model where the universe "breathes" through structured energy flow.

1. Expansion and Contraction as Part of a Larger Energy Flow

- The universe does not simply "expand forever" or "collapse"—it follows structured oscillations based on energy balance.

- Energy fluctuations within the UEM may cause localized contractions and expansions over cosmic timescales.
- This aligns with new discoveries showing galaxies maintaining density rather than uniformly dispersing.

2. Cosmic Structures as Evidence of Energy Cycles

- Large-scale structures, like the cosmic web, follow fractal-like energy distributions, suggesting cyclic self-organization.
- If expansion were purely geometric, we would expect structures to stretch apart uniformly, but they do not.
- This suggests that energy remains structured, following a regulated equilibrium cycle rather than random expansion.

❖ **UED Perspective:** The universe does not expand like a balloon but "breathes" as part of a self-sustaining energy cycle, ensuring equilibrium over cosmic timescales.

Expansion as an Energy Redistribution Process

From a UED perspective, cosmic expansion is not caused by space "stretching" but by structured energy redistribution within the UEM.

- Expansion is not a mysterious force—it is an effect of large-scale energy equilibrium adjustments.
- Dark energy is unnecessary—expansion follows natural energy redistribution rather than requiring an external force.
- The universe will not expand forever or collapse—it follows a self-regulating energy cycle, ensuring long-term balance.
- Instead of stretching infinitely, the universe "breathes" through structured energy flows within the UEM.

This leads to a profound realization:
If cosmic expansion is an effect of energy redistribution, could black holes be part of the same system? Are they not singularities but structured energy reservoirs playing a role in cosmic equilibrium?
In the next section, we explore how UED redefines black holes—not as infinite singularities but as energy storage and recycling zones that regulate energy balance in the universe.

7.6 Black Holes – Energy Storage and Recycling Zones, Not Singularities

Black holes have long been treated as regions where matter collapses into infinite density, creating singularities where space-time itself ceases to function. According to general relativity, black holes compress all matter into a point of zero volume and infinite gravity, making them paradoxical objects where known physics breaks down.

However, this view raises critical issues:

- Does space-time cease to exist inside black holes, or does it transition into a different structured energy form?
- If black holes "destroy" information and matter, how does the universe maintain its energy balance?
- Are black holes endpoints of physics, or do they serve a functional role in cosmic energy equilibrium?

From a Universal Energy Dynamics (UED) perspective, black holes are not destructive singularities but high-energy storage and recycling zones. Instead of being places where physics breaks down, black holes play a crucial role in the structured transformation of matter and energy, maintaining universal balance.

This section explores:

- Why general relativity treats black holes as singularities with infinite density.
- UED's alternative: Black holes as high-energy storage and recycling systems, preserving total energy balance.
- How black holes absorb, transform, and redistribute energy rather than "destroying" it.
- Could black holes be key mechanisms for universal energy equilibrium rather than endpoints of physics?

Why General Relativity Treats Black Holes as Singularities with Infinite Density

1. The Singularity Problem in General Relativity

- Einstein's field equations predict that when a massive star collapses, its gravity becomes so strong that space-time curves infinitely, forming a singularity.
- This implies that density and gravitational forces become infinite, creating a breakdown in known physics.

- However, infinite density is physically meaningless—it suggests that our understanding of energy transformation is incomplete.

2. The Event Horizon as a One-Way Energy Sink

- The event horizon is the boundary beyond which nothing can escape a black hole.
- Traditional physics assumes that once energy or matter crosses this threshold, it is lost forever, trapped inside the singularity.
- This creates the information paradox—if energy is truly lost, it violates the laws of conservation.

❖ **UED Perspective:** Rather than infinite compression, black holes transition matter into a stable energy state, ensuring universal energy balance.

UED's Alternative: Black Holes as High-Energy Storage and Recycling Systems

From a UED perspective, black holes do not compress matter into nothingness—they transform it into stable energy within the Universal Energy Medium (UEM).

1. Space-Time Ceases to Exist as Matter Becomes Stable Energy

- Space-time is not a fundamental entity—it is a property of mass-energy interactions.
- Once matter enters a black hole, its mass-energy is redistributed in such a way that space-time itself is no longer meaningful.
- This means that inside a black hole, space-time does not "continue"—it transitions into a different energy configuration.

2. Black Holes as Energy Processors, Not Destructive Voids

- Rather than "destroying" information, black holes function as structured energy reservoirs.
- The mass that enters a black hole is not infinitely compressed—it is stored and later redistributed.
- Instead of viewing black holes as cosmic "dead ends," they should be seen as functional components of energy flow in the universe.

❖ **UED Perspective:** Black holes do not erase information or matter—they convert them into stable energy, ensuring cosmic balance.

How Black Holes Absorb, Transform, and Redistribute Energy Rather Than "Destroying" It

1. Hawking Radiation – The Slow Energy Release from Black Holes

- Stephen Hawking proposed that black holes slowly emit radiation due to quantum effects near the event horizon.
- This means that black holes gradually lose mass over time, releasing energy back into the universe.
- If energy can escape, this challenges the idea that black holes are absolute one-way traps.

2. Energy Recycling Through Accretion Disks and Jet Ejections

- Black holes are often surrounded by accretion disks—high-energy matter that emits intense radiation before being absorbed.
- Many black holes produce relativistic jets—powerful streams of energy ejected from their poles at nearly the speed of light.
- These mechanisms suggest that black holes do not just absorb—they actively redistribute energy, playing a role in cosmic energy balance.

3. The Continuous Formation of Matter as the Counterpart to Black Hole Dissolution

- If black holes convert mass into stable energy, then the formation of new matter must be a complementary process.
- Matter formation does not happen chaotically—it follows a structured, gradual process where energy coagulates into stars, planets, and galaxies.
- This suggests that matter and energy are in continuous transition, forming and dissolving in structured cycles.

❖ **UED Perspective**: Black holes are not one-way energy sinks—they are structured energy regulators that ensure energy is continuously recycled into different forms.

Could Black Holes Be Key Mechanisms for Universal Energy Equilibrium Rather Than Endpoints of Physics?

If black holes are energy processors, they must play a vital role in maintaining the structured balance of the universe.

1. Do Black Holes Prevent Excessive Matter Accumulation?

- Without black holes, matter could accumulate in unstable configurations, leading to chaotic cosmic structures.
- Black holes act as regulators, ensuring energy remains balanced rather than allowing uncontrolled mass growth.

2. Do Black Holes Play a Role in Cosmic Expansion?

- If black holes recycle energy, could they be actively contributing to the large-scale energy distribution that drives cosmic expansion?
- Instead of seeing expansion as a one-way process, could black holes be part of a cycle where energy is stored, transformed, and later released?

3. Could Black Holes Connect to Other Energy Structures?

- Some theories propose that black holes could be linked to white holes—regions where energy is re-emitted into the universe.
- UED suggests that rather than black holes being isolated voids, they are part of a larger energy network that cycles energy across cosmic structures.

❖ **UED Perspective:** Black holes are critical to universal energy equilibrium, ensuring that energy remains structured rather than accumulating or dispersing randomly.

Black Holes as Energy Regulators, Not Singularities

From a UED perspective, black holes are not singularities—they are structured energy storage and recycling systems that maintain cosmic balance.

- Space-time ceases to exist inside black holes as matter transitions into pure stable energy.
- Black holes do not infinitely compress matter—they store, transform, and later redistribute energy.
- Processes like Hawking radiation, accretion disks, and jet ejections suggest that black holes regulate cosmic energy rather than simply "consuming" it.
- Instead of being "endpoints" of physics, black holes are essential for maintaining the energy cycle of the universe.

This leads to a profound realization:
If black holes do not "end" physics but regulate cosmic energy flow, then could the Big Bang itself be reinterpreted as an energy phase transition rather than a singular beginning?

In the next section, we explore how UED redefines the Big Bang—not as an explosion from nothing but as a structured transformation within the Universal Energy Medium, leading to the formation of the observable universe.

7.7 The Big Bang – Was It an Energy Transition Rather Than a Singular Beginning?

The Big Bang theory is the dominant cosmological model describing the origin of the universe. According to this model, the universe began as a hot, dense singularity approximately 13.8 billion years ago, expanding rapidly into what we observe today. However, this explanation raises critical unanswered questions:

- What triggered the Big Bang? If time and space began at this singularity, what existed before it?
- Did the universe truly emerge from "nothing"? This violates energy conservation laws.
- Is the Big Bang a one-time event, or could it be part of a larger cycle of energy transformations?

From a Universal Energy Dynamics (UED) perspective, the Big Bang was not an explosion from nothing but a structured energy phase transition. Instead of treating it as a singular event, UED proposes that the observable universe emerged from a prior structured energy state within the Universal Energy Medium (UEM).

This section explores:

- The standard Big Bang model assumes a singular explosive event—but what triggered it?
- UED's interpretation: The universe did not emerge from nothing—it transitioned from a prior structured energy state.
- Could the Big Bang be one phase in an ongoing cycle of energy transformations?
- If energy is always conserved, can the universe truly have a starting point?

The Standard Big Bang Model Assumes a Singular Explosive Event—But What Triggered It?

The Big Bang model explains many cosmological observations, but it also contains several theoretical inconsistencies.

1. The Singularity Problem: An Undefined Beginning

- The traditional Big Bang model states that the universe began as a singularity—an infinitely dense, infinitely small point.

- This singularity contains all space, time, and energy, yet its existence cannot be explained by known physics.
- General relativity breaks down at this singularity, meaning we have no mathematical description of what came before.

2. The Violation of Energy Conservation

- The First Law of Thermodynamics states that energy cannot be created or destroyed, only transformed.
- If the Big Bang created all energy in the universe, this would violate fundamental physical laws.
- Instead, energy must have already existed in another form before transitioning into the observable universe.

3. The Horizon and Flatness Problems: Fine-Tuned Expansion

- The universe appears smooth and uniform across vast distances, despite regions having no causal connection.
- The expansion rate is fine-tuned to prevent immediate collapse or uncontrolled dispersal.
- These issues suggest that the Big Bang was not a chaotic explosion but a structured energy process.

❖ **UED Perspective:** The universe did not begin as a singularity—it transitioned from a prior structured energy state, following conservation laws.

UED's Interpretation: The Universe Did Not Emerge from Nothing—It Transitioned from a Prior Structured Energy State

Rather than assuming a singular origin, UED proposes that the Big Bang was an energy phase transition within the Universal Energy Medium (UEM).

1. The Universe as an Evolving Energy System

- Energy always exists in structured forms, shifting between states rather than appearing from nothing.
- The "beginning" of the universe was not a creation event—it was a transformation from a previous energy configuration.

- Just as matter forms and dissolves into energy, the universe itself may cycle between different structured energy states.

2. The Big Bang as an Energy Condensation Process

- Before the Big Bang, the universe existed in a prior high-energy equilibrium state within the UEM.
- A phase transition occurred, leading to the rapid expansion and formation of matter.
- This means the Big Bang was not an explosion, but an organized restructuring of energy into a new cosmic state.

3. No Absolute Beginning—Only Energy Transformation

- If energy is always conserved, then the universe cannot have a true "starting point."
- Instead, what we call the Big Bang is just the beginning of our observable phase of universal energy evolution.

❖ **UED Perspective:** The Big Bang was not a singular origin—it was a transition from a prior structured energy equilibrium.

Could the Big Bang Be One Phase in an Ongoing Cycle of Energy Transformations?

If the universe did not begin from nothing, then it must be part of a continuous energy cycle.

1. The Universe as a Cyclic Energy System

- Some cosmological models already propose that the universe may follow an eternal cycle of expansion and contraction.
- UED refines this idea by stating that energy does not collapse into a singularity, but shifts between structured configurations.
- This means that each "Big Bang" is simply the next phase in a larger cosmic energy cycle.

2. Expansion and Contraction as Balanced Energy Flows

- Instead of the universe expanding forever or collapsing completely, UED proposes a structured cycle of expansion, stabilization, and energy reconfiguration.
- This aligns with evidence that suggests the universe is not expanding chaotically but follows structured density patterns.

- Just as black holes regulate energy within galaxies, the universe as a whole may follow a larger energy transformation cycle.

3. The Observable Universe as One Layer in a Larger Energy System

- If the Big Bang was a phase transition, what existed before it?
- UED suggests that the universe we observe is just one structured energy layer within the larger UEM.
- This means that other energy states may exist beyond our observable limits, influencing cosmic evolution.

❖ **UED Perspective:** The universe is not a one-time event—it follows structured energy cycles, ensuring continuity rather than absolute beginnings.

If Energy Is Always Conserved, Can the Universe Truly Have a Starting Point?

Energy cannot be created or destroyed, so where did the energy of the universe come from?

1. The Flaw in "Creation from Nothing" Theories

- The idea that the universe "came from nothing" contradicts the fundamental law of energy conservation.
- UED proposes that energy always existed, shifting between structured states.
- What we call the Big Bang is just the moment our observable energy structure emerged, not the true beginning.

2. The UEM as the Eternal Framework of Energy Balance

- The Universal Energy Medium (UEM) is the structured energy field in which all cosmic processes take place.
- Before the Big Bang, the universe was simply in another state of equilibrium.
- Energy is not created or destroyed—it continuously reorganizes itself through structured transitions.

❖ **UED Perspective:** The universe has no absolute beginning—only structured energy transformations within the eternal UEM.

The Big Bang as an Energy Phase Transition, Not a Singular Beginning

From a UED perspective, the Big Bang was not a singular explosion but a structured transformation of energy within the UEM.

- The Big Bang was not the start of everything—it was an energy phase transition.
- Energy conservation laws require that the universe existed in some prior state before transitioning into its current phase.
- Rather than being a unique event, the Big Bang is likely one phase in a continuous energy cycle.
- Instead of emerging from nothing, the universe evolved from a structured energy equilibrium.

This leads to a profound realization:
If the universe continuously transitions through structured energy phases, does this mean its fate is also governed by energy equilibrium rather than an ultimate "end"?
In the next section, we explore the fate of the universe—whether it will expand forever, collapse, or follow a structured cycle of energy reconfiguration, ensuring its eternal evolution.

7.8 The Fate of the Universe – A Cyclic, Self-Regulating Energy System?

The ultimate fate of the universe is one of the most profound questions in cosmology. Traditional physics presents several possibilities, ranging from eternal expansion leading to heat death to eventual contraction in a Big Crunch. However, these models often assume that the universe is a one-time event, bound by linear time and irreversible entropy.
From a Universal Energy Dynamics (UED) perspective, the universe is not on a path to an ultimate "end"—it follows structured energy cycles that maintain equilibrium over cosmic timescales. Instead of viewing the universe as an isolated system that will eventually run out of usable energy, UED suggests that the cosmos is a self-regulating energy network, where expansion, contraction, transformation, and restructuring occur in balance.
This section explores:

- What does UED predict about the long-term evolution of the cosmos?
- Is the universe headed toward heat death, or does energy continuously cycle?
- Could the cosmos regenerate itself through energy restructuring rather than ending in entropy?
- How does UED reshape our understanding of the universe's purpose and long-term sustainability?

What Does UED Predict About the Long-Term Evolution of the Cosmos?

Cosmological models often assume that the universe began at the Big Bang and will evolve toward an ultimate fate determined by expansion, gravity, and entropy. However, if the universe is fundamentally governed by structured energy transformations, then its long-term evolution must follow patterns of balance rather than linear decay.

1. The Universe as an Adaptive Energy System

- The universe is not a machine running out of fuel—it is a structured energy system that evolves dynamically over time.
- Energy does not disappear but transforms into different states, ensuring that the cosmic energy balance is maintained.
- If the universe has persisted for billions of years, why assume it will suddenly "end" rather than continue cycling?

2. Expansion, Contraction, and Energy Redistribution

- Expansion does not mean the universe is spreading into nothingness—it is an effect of energy redistribution across the Universal Energy Medium (UEM).
- As energy moves through different cosmic structures, it may cause regions of expansion and contraction without leading to total dissipation.
- This suggests that the universe may experience oscillations of energy flow rather than an irreversible fate.

❖ **UED Perspective:** The universe is not following a one-way path toward destruction—it is part of a continuous energy equilibrium that evolves dynamically.

Is the Universe Headed Toward Heat Death, or Does Energy Continuously Cycle?

The heat death scenario proposes that the universe will expand until all energy is evenly distributed, preventing new structures from forming. This would result in a cold, lifeless universe where no useful energy remains. However, UED suggests a different outcome.

1. The Problem with the Heat Death Model

- The heat death scenario assumes that entropy increases indefinitely, leading to maximum disorder.
- However, entropy does not mean "absolute destruction"—it simply describes how energy moves toward equilibrium.

- If the universe follows structured energy flow, then new energy configurations should continuously emerge rather than leading to absolute stasis.

2. How Energy May Continue Cycling Beyond Heat Death

- In a purely closed system, energy may eventually reach equilibrium and stop doing work.
- However, the universe is not a closed system in classical terms—it is an interconnected energy field within the UEM.
- This suggests that energy never stops—it simply reorganizes into new structures through large-scale cosmic transformations.

❖ **UED Perspective:** Instead of a final "death," the universe undergoes structured transitions where energy is continuously redistributed, preventing total stagnation.

Could the Cosmos Regenerate Itself Through Energy Restructuring Rather Than Ending in Entropy?

If energy is always conserved and follows structured transformation cycles, then the universe must have a natural way of regenerating itself.

1. Black Holes as Energy Recycling Centers

- Black holes do not destroy energy—they convert matter into highly compressed energy states.
- Over time, energy may be released through Hawking radiation, relativistic jets, or through other cosmic processes we have yet to fully understand.
- This suggests that black holes play a role in preventing energy stagnation, ensuring that energy remains active in the universe.

2. Cosmic Cycles: Expansion, Reconfiguration, and Renewal

- If energy follows structured transformations, then the universe may undergo continuous phases of expansion, stabilization, and reorganization.
- Instead of a single "Big Bang," the universe could experience multiple energy transitions over cosmic timescales.
- This aligns with observations of self-organizing cosmic structures, such as galaxies maintaining density patterns despite expansion.

3. UEM as the Framework for Eternal Energy Renewal

- The Universal Energy Medium (UEM) provides a structured field where energy can continuously flow and reorganize.
- Instead of a universe running out of usable energy, energy transitions into different forms, ensuring long-term sustainability.
- This means that rather than facing an "end," the universe evolves as part of an eternal cycle of energy balance.

❖ **UED Perspective:** The cosmos does not "end"—it transforms, ensuring that energy remains active and structured rather than dissipating into disorder.

How Does UED Reshape Our Understanding of the Universe's Purpose and Long-Term Sustainability?

Traditional cosmology often views the universe as a system with a beginning and an end, treating it as a one-time event rather than a continuous process. However, UED proposes that the universe:

- Is a structured energy system that self-regulates rather than following a linear path toward extinction.
- Does not require an external force to sustain it—energy transformation cycles naturally maintain equilibrium.
- Ensures sustainability through structured energy redistribution, preventing infinite expansion or total collapse.

This suggests that the universe is not an accident or a temporary event—it is a structured, evolving system that follows natural laws of energy conservation and balance.

❖ **UED Perspective:** The universe has no ultimate "end"—it is an adaptive, self-regulating energy system that continues evolving rather than collapsing or becoming stagnant.

The Universe as a Self-Sustaining Energy System

From a UED perspective, the fate of the universe is not predetermined by heat death, a Big Crunch, or infinite expansion. Instead, it follows structured energy cycles that ensure its long-term balance.

- The universe does not follow a linear path toward extinction—it evolves through structured energy transformations.
- Heat death is not inevitable—energy continuously cycles through self-regulating processes.
- Cosmic energy restructuring ensures that the universe regenerates itself rather than

collapsing or stagnating.
- Rather than an isolated system, the universe is an interconnected energy network that maintains equilibrium across time and space.

This leads to a profound realization:
If the universe is an eternal, structured energy system, does this mean that intelligence and consciousness also follow an interconnected, self-sustaining energy process? Could awareness itself be part of the universe's structured energy flow?
In the next part of this book, we move beyond physics and into the nature of intelligence, awareness, and life—exploring how UED extends beyond the cosmos into consciousness and the evolution of thought.

7.9 Conclusion: The Universe as an Interconnected Energy Network

Through our exploration of Universal Energy Dynamics (UED), we have uncovered a fundamentally new perspective on the structure, evolution, and nature of the universe. Instead of treating the cosmos as a collection of disconnected forces and events, UED reveals that the universe is an interconnected energy network, where everything follows structured energy transformations and equilibrium.
Rather than viewing space as empty, dark matter as an invisible substance, dark energy as a mysterious force, or black holes as destructive singularities, UED shows that these phenomena are all manifestations of a structured energy balance that ensures the stability of the universe. This perspective resolves long-standing paradoxes in modern physics and presents a self-sustaining, unified model for cosmic evolution.
This chapter has demonstrated several key insights:

- The universe is not empty—it operates as an energy medium that maintains equilibrium.
- Dark matter, dark energy, expansion, and black holes are manifestations of structured energy balance.
- UED provides a unified explanation for cosmic origins, structure, and evolution, resolving paradoxes in modern physics.
- Rather than randomness, the universe follows structured energy transformations, ensuring its long-term stability.

The Universe is Not Empty—It Operates as an Energy Medium, Maintaining Equilibrium

For centuries, classical physics assumed that space was a void in which celestial bodies moved. However, modern discoveries—such as quantum fluctuations, vacuum energy, and dark matter-like effects—suggest that space is not "nothing" but an active energy medium that structures interactions across cosmic scales.

1. The Universal Energy Medium (UEM) as the Foundation of the Cosmos

- The UEM fills all of space, ensuring that energy is structured rather than dispersed into disorder.
- Space itself acts as an energy-balancing system, influencing how forces, particles, and cosmic structures interact.
- This explains why the universe does not behave randomly or chaotically—energy transformations follow a structured equilibrium.

❖ **UED Perspective:** Rather than viewing space as an empty void, we must recognize it as an active energy field that governs all cosmic interactions.

Dark Matter, Dark Energy, Expansion, and Black Holes Are Manifestations of Structured Energy Balance

Many unexplained cosmic phenomena can now be understood as different expressions of energy equilibrium within the UEM.

1. Dark Matter as an Energy Network, Not an Unknown Particle

- Dark matter effects arise not from invisible particles but from structured energy maintaining gravitational balance.
- There is no need to introduce undetectable matter—the gravitational anomalies observed in galaxies are energy equilibrium effects within the UEM.

2. Dark Energy as Large-Scale Energy Redistribution

- The expansion of the universe does not require an external force—it is an effect of structured energy flow within the cosmic medium.
- What we call "dark energy" is simply the large-scale energy balance within the UEM, regulating cosmic expansion over time.

3. Black Holes as Energy Storage and Recycling Centers

- Instead of being infinite singularities, black holes serve as structured energy processors, ensuring equilibrium rather than destruction.

- Black holes do not permanently trap energy—they transform and redistribute it, playing a key role in energy conservation across cosmic structures.

❖ **UED Perspective:** These cosmic mysteries are not separate problems—they are interconnected manifestations of structured energy flow.

UED Provides a Unified Explanation for Cosmic Origins, Structure, and Evolution, Resolving Paradoxes in Modern Physics

Rather than relying on separate, sometimes contradictory models to explain different aspects of the universe, UED provides a unified framework that integrates all known physics under energy conservation and structured transformations.

1. The Big Bang as an Energy Phase Transition, Not a Creation Event

- Instead of emerging from nothing, the universe transitioned from a prior structured energy state into its current configuration.
- This eliminates the paradox of something coming from nothing, ensuring that energy conservation remains intact.

2. Expansion as a Cyclic Energy Process, Not an Infinite Growth Model

- Instead of assuming eternal expansion or collapse, UED suggests the universe cycles through structured energy transformations over cosmic timescales.
- This means the universe is not moving toward heat death—it is evolving in a self-regulating cycle.

3. The Fate of the Universe as a Self-Sustaining Energy Network

- The universe will not end in total entropy or infinite expansion—it will continue evolving through energy restructuring, preventing stagnation.
- This means that rather than a beginning and an end, the universe follows a structured process of transformation and regeneration.

❖ **UED Perspective:** Instead of viewing the universe as a system destined for collapse or endless expansion, we must see it as an interconnected, self-sustaining energy system.

Rather Than Randomness, the Universe Follows Structured Energy Transformations, Ensuring Its Long-Term Stability

One of the most profound conclusions from UED is that the universe is not a chaotic system but a self-regulating, structured energy network.

1. The Universe as a Coherent, Predictable Energy System

- Rather than relying on probabilistic randomness, UED shows that energy follows structured transformation pathways, ensuring that order is maintained.
- The evolution of cosmic structures—from galaxies to black holes to quantum particles—follows predictable energy equilibrium principles.

2. Energy Conservation as the Ultimate Cosmic Rule

- Energy is never created or destroyed—it is always structured, balanced, and recycled in different forms.
- This ensures that the universe remains stable over long periods, preventing the possibility of uncontrolled chaos or collapse.

❖ **UED Perspective:** Instead of assuming the universe is governed by randomness, we must recognize that structured energy flow is the foundation of all cosmic evolution.

Final Realization: The Universe as a Living, Self-Sustaining Energy Network

The conclusions of UED suggest a profound new way of viewing the universe—not as a mechanical system moving toward an inevitable end, but as a self-regulating, interconnected energy network that continuously evolves and maintains equilibrium.

- Space is not empty—it is an energy medium ensuring stability across all cosmic scales.
- Dark matter, dark energy, expansion, and black holes are part of an integrated energy system, not separate mysteries.
- The Big Bang was not a singular event—it was a structured phase transition within the Universal Energy Medium (UEM).
- The universe does not move toward randomness—it follows structured energy transformations, ensuring balance and sustainability.

This leads to a profound realization:

Does Intelligence and Consciousness Follow the Same Energy Equilibrium?

If the universe follows structured energy equilibrium, does intelligence and consciousness also emerge from this same interconnected energy system?

- Could thought, awareness, and life itself be part of the universe's structured energy flow?
- Could intelligence evolve as a natural consequence of structured energy complexity, rather than being unique to biological organisms?

Moving Forward: From Cosmic Energy to Intelligence and Consciousness

Having now established the energy foundation of physics and cosmology, we are ready to take the next step—exploring the role of intelligence, thought, and consciousness as structured energy interactions within UED.

In the next part of this book, we move beyond physics and into the nature of life, intelligence, and awareness—exploring how UED extends beyond the cosmos into cognition, perception, and the evolution of thought.

Answers to Big Questions

(FAQ for Part 2)

Part 2 of The Blueprint explored how Universal Energy Dynamics (UED) integrates classical and modern physics, quantum mechanics, relativity, thermodynamics, and the structure of the universe. This section provides answers to the biggest unresolved questions in modern physics, offering redefinitions of fundamental concepts such as Newton's Laws, force, momentum, light, gravity, black holes, the Big Bang, and the fate of the universe—all through the lens of UED.

1. Does UED challenge existing scientific theories like relativity and quantum mechanics?

No, UED does not reject existing theories—it integrates them into a single framework that explains how they emerge from structured energy balance.

- Classical mechanics, relativity, and quantum mechanics are not separate but interconnected through energy transformations.
- Instead of treating different scientific theories as independent, UED shows that they follow universal energy principles.

2. How does UED redefine Newton's Laws in terms of energy equilibrium?

Redefinition of Newton's Laws in UED:

First Law (Inertia as Energy Equilibrium)

- **Classical Definition:** An object at rest remains at rest, and an object in motion remains in uniform motion unless acted upon by an external force.
- **UED Definition:** An energy system in equilibrium will remain in its current state of energy distribution unless an external energy transformation disrupts its balance.

Second Law (Force as Structured Energy Transfer)

- **Classical Definition:** The force acting on an object is equal to its mass times acceleration (F=ma).
- **UED Definition:** Force is the rate at which energy redistributes within a system, proportional to the resistance (mass-energy structure) and the change in motion (acceleration).

Third Law (Action-Reaction as Energy Conservation)

- **Classical Definition:** For every action, there is an equal and opposite reaction.
- **UED Definition:** Every energy transfer must result in a counterbalancing redistribution, ensuring that total system energy remains in equilibrium.

3. How does UED redefine force and momentum?

Redefinition of Force and Momentum in UED:
Force in UED

- **Classical Definition:** Force is the rate of change of momentum (F=ma).
- **UED Definition:**Force is the rate of energy redistribution that alters the structured motion of a system, adjusting its equilibrium state.

Momentum in UED

- **Classical Definition:** Momentum (p) is the product of mass and velocity (p=mv).
- **UED Definition:** Momentum is the structured flow of energy contained within a moving system, defined by its energy density (mass) and its directed transformation (velocity).

4. How does UED redefine the propagation of light and photons?

Redefinition of Light Propagation in UED:
Classical View: Light is an electromagnetic wave that propagates through space as a massless particle (photon) and a wave simultaneously.
UED Perspective:

- Light is a structured energy transfer mechanism, where photons act as energy messengers across different realms.
- Light does not "travel" through space—it propagates as a structured energy wave within the

Universal Energy Medium (UEM).
- The speed of light (c) is not an absolute limit—it is the rate at which energy stabilizes within the medium.
- Photons are not "particles" in the traditional sense but structured energy packets that relay information between systems.

5. How does UED explain black holes beyond singularities?

Traditional physics treats black holes as infinite-density singularities where space-time collapses. However, UED suggests that:

- Black holes are not singularities but high-energy storage and recycling zones.
- Matter does not "disappear" inside a black hole—it transitions into stable energy, ceasing to interact with space-time.
- Black holes do not violate energy conservation—they regulate energy balance in the universe.
- Rather than being endpoints, black holes are part of a structured energy flow, ensuring cosmic stability.

6. Was the Big Bang truly the beginning of the universe?

The Big Bang model suggests that the universe began as a singularity, but this leads to paradoxes:

- If energy cannot be created or destroyed, how could the universe emerge from nothing?
- What existed before the Big Bang?

UED Perspective:

- The Big Bang was not the beginning—it was an energy phase transition within the Universal Energy Medium (UEM).
- The universe did not "emerge" from nothing—it evolved from a prior structured energy state.
- The Big Bang was not an explosion—it was an organized restructuring of energy into a new cosmic phase.

7. What does UED say about gravitational waves?

Gravitational waves are traditionally understood as ripples in space-time caused by accelerating massive objects. However, UED suggests that:

- Gravitational waves are not space-time distortions but structured energy redistributions.
- Instead of traveling through "empty" space, gravitational waves propagate through the UEM, balancing mass-energy fluctuations.
- This means that gravity is not a separate force but a structured energy interaction maintaining cosmic stability.

8. Does the universe have an ultimate end, or is it self-sustaining?

Traditional models propose three possible fates:

- **Heat Death** – The universe expands until all energy is evenly distributed, making work impossible.
- **Big Crunch** – The universe collapses back into a singularity.
- **Big Rip** – The universe expands infinitely, tearing apart all structures.

UED Perspective:

- The universe does not "end"—it follows structured cycles of energy transformation.
- Instead of permanent heat death, energy continues cycling through expansion, contraction, and restructuring.
- Black holes, expansion, and cosmic energy flows act as regulators, preventing infinite collapse or total dissipation.
- The universe is a self-regulating energy network, continuously evolving rather than reaching a terminal fate.

Final Thoughts: Where Do We Go From Here?

Having now established that physics, quantum mechanics, and cosmic evolution are all governed by structured energy equilibrium, the next phase of our exploration leads us to intelligence, consciousness, and perception.

- If structured energy explains physics, can it also explain life and intelligence?

- Does consciousness emerge from structured energy interactions, similar to entanglement?
- Could awareness, decision-making, and intuition be manifestations of universal energy balance?

Schrödinger's Cat and UED: Does Observation Define Reality?

Schrödinger's cat experiment is often interpreted as evidence that quantum states remain undefined until observed. It suggests that until measured, a cat inside a box with a quantum-triggered poison is both alive and dead simultaneously.

However, from a Universal Energy Dynamics (UED) perspective, this is not a paradox—it is a misunderstanding of how energy equilibrium determines quantum states.

1. The Traditional View: Does Observation Create Reality?

- Quantum mechanics suggests that a system exists in multiple states (superposition) until measured.
- Schrödinger's cat experiment implies that observation "collapses" reality into one definite outcome.
- This raises the paradox: Does measurement create reality, or does reality exist independently of observation?

2. UED's Explanation: Superposition as an Energy Equilibrium State

- The cat is never in two states at once—its energy state is structured before measurement occurs.
- "Wave function collapse" is actually an energy redistribution process that follows equilibrium laws, not an arbitrary selection.
- This means the cat is already in one state before observation, but the observer's knowledge of that state is limited.

3. Measurement Does Not Create Reality—It Reveals Pre-Existing Energy Equilibrium

- The idea that observation "collapses" reality is flawed—energy structures already exist in a stable configuration.
- Measurement is not a force that determines an outcome—it is simply accessing an already balanced energy system.
- Quantum superposition is not about reality being undefined—it is about structured energy awaiting transition.

4. The Role of Entanglement in Resolving the Paradox

- The cat's fate is already entangled with the energy system inside the box.
- The outcome is determined by structured energy interactions, not by human observation.
- This aligns with UED's principle that quantum mechanics is not random but follows structured energy pathways.

5. Conclusion: Schrödinger's Cat Is an Energy Balance Question, Not an Observer Effect

- From a UED perspective, the Schrödinger's cat paradox disappears when we recognize that:
- The cat is never both alive and dead—it follows structured energy equilibrium.
- Observation does not "collapse" reality—it reveals an energy balance that already existed.
- Superposition is not a paradox—it is a structured energy transformation awaiting resolution within equilibrium laws.
- Entanglement plays a role, but it follows structured energy pathways rather than creating random collapses.

Final UED Interpretation:
Schrödinger's cat experiment is not proof that reality is undefined—it is proof that energy balance determines all quantum outcomes long before observation occurs.

Energy in Motion – The Transition Between Active and Stable States

Throughout Part 2, we explored Universal Energy Dynamics (UED) and its role in explaining motion, force, gravity, expansion, black holes, and the structure of the universe. Now, we refine our understanding further by recognizing that all energy transitions—whether in matter, space, or consciousness—occur through structured motion, maintaining universal equilibrium.

A key realization is that energy never remains stagnant. It continuously moves, interacts, and transitions between two fundamental states:

- **Active Energy** – Energy in motion, transformation, and interaction.
- **Stable Energy** – Energy in a perfectly balanced, infinite state, beyond structured interactions.

This transition between Active and Stable Energy governs all cosmic, physical, and intelligent processes and ensures the conservation of energy across the universe.

1. The Two Fundamental States of Energy in UED

Active Energy (Energy in Motion, Transformation, and Interaction)

Energy that moves, drives interactions, enables transformations, and ensures equilibrium across systems.

- Found in kinetic energy, radiation, heat transfer, mechanical motion, and thought processes.
- Appears in cosmic expansion, particle motion, gravitational waves, and consciousness.
- In living beings, it manifests as awareness, decision-making, and the ability to direct energy flow.
- In matter, it appears as atomic vibrations, phase transitions, and chemical interactions.

Stable Energy (The Infinite State, Where Matter Loses Form)

Energy that exists in its purest, most fundamental state, where all structured forms dissolve into equilibrium.

- Found in black holes, the pre-Big Bang state, and the vacuum structure of space.
- Matter loses all identity in this state—it no longer exists as mass, charge, or particles.
- This is not destruction or chaos—it is the ultimate energy equilibrium, ensuring conservation.

- It is the source and return point of all structured energy across the universe.

Key UED Principle:
The universe cycles between Active Energy (structured motion and transformation) and Stable Energy (perfect equilibrium), ensuring continuity and conservation.

2. Energy Transfer in Matter – Consciousness as the Enabler of Energy Transition

- In matter, motion exists at the particle level, ensuring that energy is never static.
- Consciousness manifests as structured motion, allowing energy to move and interact.
- This motion appears at quantum and macroscopic scales, governing all physical interactions.
- Without this inherent motion, energy would not transfer, meaning no transformation could occur.

Key UED Principle:
Consciousness in matter is the structured process that enables energy transition, maintaining energy balance across systems.

Examples of Energy Transition in Matter

Process	Energy Mechanism	Type of Transition
Heat Transfer	Particles absorb thermal energy, increasing kinetic motion.	Active Energy Transition
Mechanical Motion	Force application causes displacement or deformation.	Energy Redistribution
Chemical Reactions	Molecular bonds break and reform, changing energy states.	Structural Energy Balance

Conclusion: Consciousness in matter is energy-driven motion, ensuring all physical phenomena.

3. Energy Transfer in Stable Energy or Absolute Vacuum – Electromagnetic Waves as the Enabler

In stable energy states (absolute vacuum, deep space, or the black hole singularity), electromagnetic waves serve as the primary energy carriers.

- Since vacuum has no particles for mechanical waves to propagate, electromagnetic radiation ensures continuous energy flow.
- This proves that even in the absence of physical matter, energy is always in motion.
- Light, radio waves, cosmic microwave background radiation, and gravitational waves are all examples of energy transfers in vacuum.

Key UED Principle:
Even in absolute vacuum, energy transitions occur through wave propagation, maintaining universal energy equilibrium.

Examples of Energy Transfer in Vacuum

Process	Energy Carrier	Mechanism of Transfer
Light from the Sun	Electromagnetic Waves (Photons)	Radiation propagates through empty space.
Cosmic Microwave Background (CMB)	Microwave Radiation	Residual energy from the early universe.
Gravitational Waves	Space-time Waves	Energy propagates through cosmic fabric.

Conclusion: Energy cannot be "lost" in vacuum—it transfers through waves, ensuring equilibrium across the universe.

4. All Energy Transfers Through Waves – Mechanical or Electromagnetic

One of the most profound insights is that all energy transitions, whether in matter, vacuum, or structured equilibrium, are governed by wave-based propagation.

- In matter, mechanical waves (sound, heat, force) transfer energy through particle motion.
- In vacuum, electromagnetic waves (light, gravitational waves, quantum fluctuations) transfer energy without requiring a medium.
- Across all systems, energy propagates as structured waves, maintaining balance and ensuring conservation.

Key UED Principle:

Energy is never stagnant—it continuously redistributes through wave-based interactions, preserving universal equilibrium.

5. The First Disturbance – The Birth of Active Energy from Stable Energy

- The universe did not "begin" with the Big Bang—it transitioned from Stable Energy to Active Energy.
- The First Disturbance was the structured transformation that set energy into motion.
- Matter formation was not an explosion—it was a quantum-level condensation of Active Energy.
- Black holes are not endpoints—they are return points where Active Energy dissolves back into Stable Energy, ensuring eternal balance.

Key UED Principle:
The First Disturbance was the quantum-level activation of structured energy flow, ensuring that matter, motion, and evolution follow energy equilibrium.

6. Consciousness as the Mastery of Energy Transition

- Living beings exhibit intelligence because they can control and direct Active Energy.
- Machines function as artificial consciousness because they externally regulate energy transitions.
- The universe itself may be an ultimate intelligence, maintaining structured energy flow across time and space.

Final UED Perspective:
Consciousness is the structured process of energy moving between Active and Stable states. Intelligence is the ability to control this transition efficiently. The universe itself may be a form of ultimate intelligence, maintaining perfect energy balance across all scales.

7. Final Conclusion: The Universal Balance of Energy and Existence

- Active Energy is motion, transformation, and interaction—it appears in forces, light, gravity, and thought.

- Stable Energy is infinite balance, where matter loses all form—it appears in black holes and pre-Big Bang conditions.
- The universe does not "end"—it transitions between these two states in an eternal cycle.
- Black holes are not destruction points—they are energy stabilizers that return matter to its fundamental state.
- Consciousness is the structured transition of energy between these states, directing transformation and balance.
- The universe, as a whole, functions as an intelligent system that maintains energy equilibrium across all existence.

This leads to the next profound realization:
If energy balance governs all motion, consciousness, and cosmic cycles, does intelligence itself follow an energy-based evolution? Could the structure of thought, memory, and decision-making be energy transitions just like phase changes and forces?

PART 3: The Emergence of Life, Intelligence, and Consciousness

Chapter 8: The Energy of Life – Biology and UED

In the previous parts of The Blueprint, we explored how Universal Energy Dynamics (UED) governs the structure of the universe, ensuring that all cosmic and physical phenomena follow structured energy transformations. We established that energy never remains stagnant—it continuously moves, interacts, and transitions between Active and Stable states, maintaining equilibrium across all realms of existence.

Now, we extend this principle to biology, demonstrating that life itself is an emergent phenomenon of structured energy flow. From the smallest molecular interactions to the complexity of entire ecosystems, biological systems follow the same fundamental energy conservation laws observed in physics.

The Transition from Cosmic Evolution to Biological Evolution

- If the universe follows structured energy balance, then life must have emerged as an extension of this universal principle.
- The emergence of living systems is not a random event—it is a natural outcome of energy optimization, self-organization, and transformation.
- DNA, cellular processes, and metabolism function as energy optimization systems, ensuring survival and adaptation through structured biochemical transitions.

Thus, in this chapter, we explore:

- How biological processes obey energy conservation laws.
- Why DNA, cellular processes, and metabolism are energy optimization systems.
- How UED links to evolution, self-organization, and the continuous emergence of life.

This chapter establishes biology as a natural continuation of UED, proving that life, intelligence, and consciousness are manifestations of structured energy in action. By understanding life as an energy-driven process, we take the first step toward uncovering how intelligence and self-awareness emerge as extensions of universal energy principles.

8.1 Biology as an Energy-Driven Process – Life and Thermodynamics

Life is not a separate entity from physics—it is an energy-driven process that follows the same universal laws governing cosmic and physical phenomena. Biological systems function as structured energy optimization systems, continuously extracting, utilizing, and conserving energy to sustain existence.

At its core, life is a self-regulating mechanism that emerged to optimize energy flow within structured environments. Just as the universe cycles between Active and Stable Energy, living organisms maintain their own internal energy equilibrium, balancing growth, metabolism, adaptation, and survival.

- In this section, we explore:
- How life emerged as an energy optimization system.
- Why biological systems obey energy conservation and entropy laws.
- How living organisms regulate energy to sustain structure and function.

1. How Life Emerged as an Energy Optimization System

The emergence of life is not a coincidence—it is the natural result of energy self-organization. From the earliest molecular interactions to complex organisms, life arose as an energy-efficient system designed to extract, store, and use energy for survival.

1 The Energy Origins of Life

- Before life existed, chemical and molecular interactions followed structured energy flows, seeking stability.
- The prebiotic environment was filled with active energy sources—solar radiation, geothermal heat, lightning, and cosmic rays.
- Simple molecules absorbed energy, leading to more complex structures, just as Active Energy condenses into matter.
- The first biological molecules formed not randomly, but as a consequence of energy self-organization, optimizing molecular stability.

2 The First Self-Sustaining Systems

- Over time, certain chemical reactions became self-sustaining, creating energy loops.
- Autocatalytic reactions (where molecules facilitate their own replication) became the foundation for biological self-organization.
- These processes followed the same UED principles of energy equilibrium, ensuring energy was never wasted but continuously cycled.

Key UED Principle: Life emerged as a structured energy optimization system, ensuring that energy transformations became self-regulating rather than chaotic.

Why Biological Systems Obey Energy Conservation and Entropy Laws

The laws of thermodynamics, which govern energy in physical systems, also apply to life. Every biological process—from cellular respiration to evolution—is structured around energy conservation and entropy regulation.

1 The First Law of Thermodynamics in Biology – Energy Conservation

The First Law of Thermodynamics states that:

Energy cannot be created or destroyed; it can only be transformed from one form to another.

In biology, this principle ensures that:

- Organisms do not create energy—they extract and convert it.
- Food, sunlight, and chemical reactions serve as energy sources, which are transformed into usable biological energy.
- Cells do not "generate" energy—they convert stored energy into action through metabolism.

2 The Second Law of Thermodynamics in Biology – Entropy and Energy Flow

The Second Law of Thermodynamics states that:

Entropy (disorder) in a system will increase over time unless energy is input to maintain order.

Life appears to "defy" entropy because it maintains order, but this is only possible because it continuously absorbs and processes energy.

- Organisms must consume energy to counteract entropy and maintain structured systems (e.g., cellular function, DNA replication).
- If energy flow stops, entropy increases, leading to decay, aging, and death.
- Life does not break the second law—it follows it by requiring continuous energy intake to sustain structured complexity.

Key UED Principle: Biological systems obey energy conservation and entropy laws, ensuring that life functions within the same structured energy transformations seen in the cosmos.

How Living Organisms Regulate Energy to Sustain Structure and Function

Life is an ongoing process of energy intake, transformation, and balance. Every organism regulates energy to sustain its structure, adapt to environments, and ensure survival.

1 Energy Flow in Living Systems – From Input to Output

All living systems operate through energy input, processing, and regulated output.

- **Energy Intake:** Organisms absorb energy from the environment (food, sunlight, chemical reactions).
- **Energy Conversion:** Cells convert energy into usable forms (ATP, biochemical reactions).
- **Energy Utilization:** Organisms use energy for movement, reproduction, healing, and adaptation.
- **Energy Output:** Waste heat and byproducts are dissipated to maintain energy equilibrium.

2 Metabolism as an Energy Optimization System

- Metabolism is not just a series of chemical reactions—it is an energy-regulating mechanism ensuring efficiency.
- ATP (adenosine triphosphate) acts as a structured energy storage and transfer system, ensuring energy is available when needed.
- Enzymes act as energy regulators, reducing energy waste and ensuring precise control of biochemical transformations.

3 Energy Homeostasis – The Self-Regulating System of Life
Life does not operate at maximum energy capacity—it operates at optimal energy balance.

- Too little energy leads to system failure (starvation, death).
- Too much energy causes instability (heat stress, metabolic overload).
- Organisms have evolved feedback systems to regulate energy flow, ensuring stable function.

Key UED Principle: Life sustains itself through energy regulation, ensuring structured energy flow, conservation, and optimization.

4. The Grand Realization: Life as an Energy Manifestation in Action

Now, we see how life fits into the grand structure of UED:

- Life did not "appear"—it emerged as an energy optimization process, ensuring structured transformations.
- Biological systems obey the same thermodynamic laws that govern the universe.
- Living organisms regulate energy to sustain structure, proving that life is an extension of structured energy flow.

Final UED Perspective:

Life is not separate from physics—it is an energy-driven process, following the same structured transformations as cosmic systems.

This leads directly into the next sections, where we explore:

- How DNA and cellular processes optimize energy for survival.
- How metabolism functions as a precise energy regulation system.
- How evolution follows energy adaptation principles.

8.2 DNA and Cellular Processes – The Blueprint of Energy Optimization

Life is an energy-driven system, and nowhere is this more evident than in DNA and cellular processes. DNA is not just the molecule of life—it is a highly efficient energy-storing and transmitting structure, ensuring that biological information and energy flow remain structured across generations.

Cells, the fundamental units of life, function as energy-processing systems, extracting, converting, and utilizing energy with maximum efficiency. Biological replication, from DNA duplication to cell division, follows precise energy optimization principles, ensuring that life continues without unnecessary energy loss.

This section explores:

- DNA as an energy-storing and transmitting structure.
- How cellular metabolism extracts, converts, and utilizes energy.
- Why biological replication is an energy-efficient process.

DNA as an Energy-Storing and Transmitting Structure

DNA is not just a carrier of genetic instructions—it is a structured energy information system that ensures the continuity of life.

1 The Energy Role of DNA – Storing and Transmitting Information Efficiently

- DNA is composed of nucleotides (adenine, thymine, cytosine, guanine), which store genetic information using minimal energy.
- The double-helix structure of DNA is an optimized energy state, requiring minimal input for stability.
- Hydrogen bonds between base pairs allow information storage using the least possible

energy, while still allowing replication when needed.

2 DNA Replication as an Energy-Efficient Process

- Enzymes like helicase and DNA polymerase do not "force" replication—they follow structured energy pathways.
- DNA's structure ensures minimal energy expenditure during duplication, reducing biological entropy.
- Energy stored in ATP is used precisely during replication, ensuring no excess energy is wasted.

Key UED Principle: DNA is not just a genetic code—it is an optimized energy structure that ensures minimal energy use for storing and transmitting biological information.

How Cellular Metabolism Extracts, Converts, and Utilizes Energy

Cells function as biological reactors, converting external energy sources (food, sunlight, chemicals) into usable biological energy. This follows UED's principles of energy conservation and optimization.

1 Energy Extraction – How Cells Capture Energy from the Environment

- Cells do not "generate" energy—they extract it from external sources like glucose, fats, or light.
- Photosynthesis (in plants) and cellular respiration (in animals) are highly structured energy conversion processes.
- No energy is wasted—cells absorb and regulate energy based on need, ensuring precise energy balance.

2 ATP – The Universal Energy Currency of Life

- ATP (Adenosine Triphosphate) acts as a highly efficient energy carrier, optimizing energy storage and transfer.
- ATP stores energy in phosphate bonds, which release precise amounts of energy upon hydrolysis.
- Cells do not store excessive ATP—they produce it only when needed, preventing energy waste.

3 Metabolic Pathways as Structured Energy Networks

- Glycolysis, the Krebs cycle, and oxidative phosphorylation are not random reactions—they follow structured energy flow patterns.
- Each step in metabolism maximizes energy output while minimizing energy loss.
- Feedback mechanisms ensure that energy production adapts to biological demand, preventing excess or deficit.

Key UED Principle: Metabolism is an energy regulation system, ensuring structured extraction, transformation, and utilization of energy to sustain life.

Why Biological Replication is an Energy-Efficient Process

Biological replication—from DNA duplication to cell division—follows structured energy principles, ensuring precision and efficiency.

3.1 DNA Replication – The Energy-Optimized Blueprint for Continuity

- DNA polymerase does not waste energy—it follows structured pathways to ensure accurate replication.
- Base pairing (A-T, C-G) minimizes energy loss by ensuring predictable and reversible bonding.
- Error-checking mechanisms prevent mutations, reducing energy spent on repairs.

3.2 Cell Division as an Energy-Regulated Process

- Mitosis and meiosis are not random—they occur only when energy conditions are optimal.
- Cells pause division if energy resources are insufficient (G1/S checkpoints), preventing unnecessary energy loss.
- Cellular structures like the mitotic spindle use ATP-efficient mechanisms to ensure precision in chromosome separation.

3.3 Protein Synthesis as an Energy-Optimized Process

- Ribosomes use messenger RNA (mRNA) as an instruction set, ensuring minimal energy expenditure per protein.
- Amino acid sequences are assembled in the most energy-efficient order to prevent unnecessary folding errors.
- Chaperone proteins assist in folding, reducing entropy and energy waste.

Key UED Principle: Biological replication, from DNA synthesis to cell division and protein production, follows structured energy principles to ensure efficiency and conservation.

The Grand Realization: Life as an Optimized Energy System

Now, we see how life is an extension of UED principles:

- DNA is an energy-efficient storage system, ensuring genetic continuity with minimal energy loss.
- Cellular metabolism extracts, transforms, and utilizes energy in structured, regulated ways.
- Biological replication follows structured energy pathways, optimizing continuity and efficiency.

Final UED Perspective:
Life is not just chemistry—it is an energy-optimized system, structured to extract, process, and conserve energy with maximum efficiency.
This leads directly into the next sections, where we explore:

- How metabolism fine-tunes energy utilization.
- How evolution follows energy adaptation mechanisms.
- How intelligence emerges as an advanced energy optimization process.

8.3 Metabolism as an Energy Transformation System

Life is a continuous process of energy extraction, transformation, and utilization. Every biological function—from movement to thought to reproduction—requires structured energy flow, ensuring efficiency, conservation, and sustainability.
Metabolism is the central process that enables this structured energy transformation. It is not just a collection of biochemical reactions; it is a highly optimized energy distribution network, ensuring that every cell, organ, and system receives energy in the right amount at the right time.
In this section, we explore:

- The role of ATP in structured energy transfers.
- How enzymes and biochemical reactions optimize energy efficiency.
- Why energy balance determines health, growth, and survival.

The Role of ATP in Structured Energy Transfers

ATP (Adenosine Triphosphate) is often called the "energy currency" of life, but it is much more than that. ATP acts as an energy bridge, facilitating controlled and precise energy transfers across biological systems.

1 ATP as an Energy Carrier

- ATP is not just a random energy molecule—it is a structured energy storage system.
- The phosphate bonds in ATP store potential energy, which is released when these bonds break (hydrolysis).
- ATP does not release all energy at once—it transfers specific amounts, ensuring precision and avoiding waste.

2 The ATP Cycle – Continuous Energy Recycling

- ATP is not stored long-term—it is constantly regenerated and used in an energy cycle.
- The body does not "store" large amounts of ATP—it maintains a balance by continuously breaking down and resynthesizing ATP as needed.
- This aligns with UED's principle of structured energy flow—energy is always in motion, never stagnant.

3 ATP in Cellular Work – The Three Types of Energy Use

Cells use ATP for three primary types of work:

- Mechanical Work – Muscle contractions, cell movement, cytoskeletal reorganization.
- Transport Work – Active transport of molecules across membranes.
- Chemical Work – Synthesis of macromolecules (proteins, DNA, etc.).

Key UED Principle: ATP does not "generate" energy—it acts as an optimized energy transfer molecule, ensuring structured energy flow in biological systems.

How Enzymes and Biochemical Reactions Optimize Energy Efficiency

Without optimization, metabolic reactions would be too slow and energy-intensive for life to sustain itself. Enzymes act as energy regulators, ensuring efficiency and precision.

1 Enzymes as Biological Catalysts – The Energy Optimization Mechanism

- Enzymes lower activation energy, allowing reactions to proceed with minimal energy expenditure.
- Each enzyme is highly specific, ensuring that no excess energy is wasted on unnecessary reactions.

- Enzymes speed up reactions while maintaining precise control over energy release.

2 The Lock-and-Key Model – Precision in Energy Use

- Enzymes do not force reactions—they guide energy flow along the most efficient path.
- The lock-and-key mechanism ensures that only the correct substrate binds to an enzyme, preventing energy waste.
- This mirrors UED's principle of structured energy transitions—energy does not disperse randomly; it follows the most efficient pathway.

3 The Role of Cofactors and Coenzymes in Energy Transfers

- Some enzymes require cofactors (metal ions) or coenzymes (vitamins) to function efficiently.
- These molecules act as intermediate energy carriers, ensuring smooth energy transitions.
- Without them, metabolism would become inefficient, leading to energy imbalances and metabolic disorders.

Key UED Principle: Enzymes act as energy gatekeepers, ensuring that biochemical reactions follow the most efficient energy pathways, preventing waste and maintaining equilibrium.

Why Energy Balance Determines Health, Growth, and Survival

A structured energy flow is critical for maintaining homeostasis, growth, and longevity. If energy input, processing, or output is disrupted, biological dysfunction occurs.

1 Energy Balance – The Foundation of Health

- Every biological system operates within a narrow range of energy balance.
- Too much energy intake leads to metabolic overload, fat accumulation, and disease.
- Too little energy intake results in cellular damage, loss of function, and system collapse.

2 Growth and Development as Energy Optimization

- Growth is not just an increase in size—it is a structured energy process where new cells integrate into an existing energy network.
- Cell division requires precise energy control—too much growth leads to cancer, too little leads to developmental defects.
- Hormonal regulation ensures that energy is allocated properly to growth, healing, and regeneration.

3 Survival Mechanisms and Energy Adaptation

- When energy intake drops, cells enter energy-saving modes (starvation response, autophagy).
- When energy demand increases, metabolism speeds up to meet requirements.
- Adaptation to stress (heat, cold, starvation) follows structured energy transitions, ensuring survival in changing environments.

Key UED Principle: Health, growth, and survival are all regulated by structured energy balance—life thrives when energy transitions follow optimal pathways.

The Grand Realization: Metabolism as a Self-Regulating Energy Network

Now, we see that metabolism is not just a collection of reactions—it is a self-regulating energy distribution system.

- ATP acts as the structured energy currency, ensuring precise transfers.
- Enzymes guide reactions along optimal energy pathways, preventing waste.
- Biological health, growth, and survival depend on maintaining structured energy balance.

Final UED Perspective:
Metabolism is a structured energy transformation system, ensuring that energy flows efficiently, maintaining biological equilibrium and survival.
This leads directly into the next sections, where we explore:

- How evolution follows energy adaptation principles.
- How intelligence emerges as an advanced energy optimization process.
- How consciousness represents structured energy flow in the highest form.

8.4 - Photosynthesis as an Energy Structuring Mechanism

Photosynthesis is one of nature's most efficient energy structuring processes, converting light energy into chemical energy through a highly optimized mechanism. It is not just a biological process but a fundamental example of structured energy transformation, demonstrating how energy flows, interacts, and reorganizes within living systems.
In the framework of Universal Energy Dynamics (UED), photosynthesis serves as a clear representation of how light energy is systematically absorbed, stored, and utilized to sustain life, just as other forms of structured energy interactions govern cosmic, quantum, and biological systems.

Energy Capture: The Role of Light in Photosynthesis

Light is the primary input energy source in photosynthesis. However, plants and photosynthetic organisms do not simply absorb light randomly; instead, they use specialized molecular structures to capture and optimize energy transfer.

- Chlorophyll and light-absorbing pigments act as structured energy absorbers.
- The absorbed photons excite electrons to higher energy states, initiating a cascade of energy transfers.
- These transfers occur in a highly structured system—light-dependent reactions ensure that energy is used efficiently rather than being lost as heat.

Key UED Insight: The way light interacts with chlorophyll is similar to the photoelectric effect in physics, where energy is absorbed and electrons are released. This structured energy transfer mechanism ensures maximum efficiency in energy retention and utilization.

Quantum Coherence in Photosynthesis: A Near-Perfect Energy Transfer Mechanism

One of the most fascinating aspects of photosynthesis is how energy moves through the system with near-perfect efficiency. Recent research has revealed that photosynthetic organisms use quantum coherence to optimize energy transfer.

- Instead of moving energy in a stepwise, inefficient manner, excitons (energy carriers) take multiple possible paths simultaneously.
- This wave-like behavior allows for the most efficient route to be chosen dynamically, reducing energy loss.
- Essentially, the energy transfer behaves like a coherent quantum system, meaning that biological systems are optimized for energy structuring at the quantum level.

Key UED Insight: This suggests that energy follows structured equilibrium principles even at microscopic scales. Just as electrons move through energy levels in atoms, light energy in photosynthesis is structured to follow the path of least resistance, ensuring efficiency.

Photosynthesis and the Concept of Energy Balance

At a broader level, photosynthesis is not just an energy absorption process—it is a key regulator of planetary energy balance.

- Plants, algae, and photosynthetic bacteria act as energy stabilizers, preventing excess solar energy from being wasted.
- The process maintains atmospheric equilibrium, ensuring that carbon dioxide and oxygen

cycles remain stable.

- Without photosynthesis, Earth's energy balance would collapse, leading to rapid shifts in atmospheric composition and global energy flow disruptions.

Key UED Insight: This aligns with the principle of universal energy balance, where energy is neither lost nor created but continuously structured into stable forms. In this case, light energy is converted into stored chemical energy, ensuring long-term equilibrium in living systems.

Photosynthesis as a Model for Energy-Based Technologies

The efficiency and structured nature of photosynthesis make it a model for future energy technologies, including:

- Artificial Photosynthesis – Developing synthetic systems that mimic plant energy absorption to create clean, sustainable energy sources.
- Quantum-Based Energy Transfer – Using coherent energy transport principles in advanced solar panels, superconductors, and bio-inspired nanotechnology.
- Energy Storage Optimization – Learning how plants store energy in chemical bonds efficiently to improve battery and fuel cell design.

Key UED Insight: If biological systems can store and use energy optimally, then energy structuring principles can be applied beyond biology to physics, engineering, and even propulsion technologies. Spacecraft and energy systems could be designed using the same structured energy optimization as photosynthesis.

The Link between Photosynthesis and Consciousness

Since energy structuring is a fundamental process in photosynthesis, could higher forms of structured energy interactions—like those in the brain—follow similar principles?

- The brain processes information through electrical signals and structured energy waves.
- Similar to how plants structure energy for metabolic functions, the brain structures neural signals for cognition and perception.
- This suggests that life itself is an optimized energy structuring phenomenon, where consciousness may emerge as a highly organized energy field interacting with the brain's structure.

Key UED Insight: If photosynthesis is a structured energy transformation process, then consciousness may be a structured information-energy interaction. This reinforces the idea that intelligence, life, and even thought processes are fundamentally governed by energy equilibrium dynamics.

Final Conclusion: Photosynthesis as a Universal Energy Optimization Mechanism

- Photosynthesis is a prime example of how nature structures energy for efficiency.
- The principles governing light absorption, energy transfer, and storage align with UED's structured energy framework.
- The way life optimizes energy interactions suggests that similar structuring principles apply to all scales of existence, including consciousness and advanced energy technologies.

8.5 Evolution as an Energy Adaptation Mechanism

Evolution is often described as a process of genetic changes over time, but at its core, it is an energy adaptation mechanism. Life does not evolve randomly; it follows structured energy efficiency principles, ensuring that each adaptation optimizes energy use for survival.

Every trait that emerges through evolution is a reflection of energy balance—organisms evolve to maximize energy efficiency, reduce waste, and enhance survival with minimal expenditure. From metabolic efficiency to structural adaptations, natural selection is a process of energy refinement.

In this section, we explore:

- How natural selection follows energy efficiency principles.
- Why evolutionary adaptations are energy-optimized survival mechanisms.
- The role of environmental energy in guiding evolutionary pathways.

How Natural Selection Follows Energy Efficiency Principles

Natural selection is fundamentally an energy-selection process. Traits that require less energy to maintain while maximizing survival benefits are favored over traits that demand excessive energy.

1 Energy Cost vs. Survival Benefit – The Core of Evolution

- Organisms with lower energy costs per function have a survival advantage.
- Traits that increase energy efficiency in movement, digestion, reproduction, and survival behaviors are naturally favored.
- Evolution does not promote complexity for its own sake—it selects traits that optimize energy use in a given environment.

2 Why Simplicity Often Wins in Evolution

- Simple structures that achieve the same function as complex ones are more energy-efficient.
- Many organisms, like bacteria, have remained unchanged for millions of years because they already operate at peak energy efficiency.
- Larger, more complex organisms must develop systems to compensate for their higher energy demands, leading to innovations in metabolism and intelligence.

Key UED Principle: Natural selection is an energy refinement process—only the most energy-efficient traits are retained over time.

Why Evolutionary Adaptations Are Energy-Optimized Survival Mechanisms

Every adaptation in nature is an energy solution to an environmental challenge. The most successful adaptations are those that allow organisms to perform essential functions using the least energy.

1 Energy-Efficient Traits in Evolution

- Aerodynamic wings in birds and insects reduce energy cost in flight.
- Camouflage in animals minimizes movement, reducing energy spent on hiding.
- Hibernation and torpor in mammals conserve energy when resources are scarce.

2 The Evolution of Metabolic Efficiency

- Early organisms relied on anaerobic respiration (low energy output).
- The emergence of aerobic respiration provided 15x more energy per glucose molecule, enabling complex life.
- This metabolic shift led to energy surplus, allowing for larger organisms, brains, and intelligence.

3 Intelligence as an Energy Optimization Strategy

- Higher intelligence evolved as a means to reduce unnecessary physical energy expenditure.
- Instead of chasing prey endlessly, intelligent predators use strategy, planning, and group coordination to optimize energy use.
- The human brain, despite its high energy demand, allows for energy-efficient problem-solving, tool use, and environmental control.

Key UED Principle: Evolution does not just select for survival—it selects for the most energy-efficient way to survive.

The Role of Environmental Energy in Guiding Evolutionary Pathways

Evolution is not just driven by random mutations—it is guided by environmental energy conditions. The availability of energy resources directly influences which traits evolve and which species thrive.

1 Energy-Rich Environments Lead to Complexity

- Ecosystems with abundant energy (sunlight, oxygen, food supply) tend to produce larger, more complex organisms.
- The Cambrian Explosion (540 million years ago) occurred when oxygen levels rose, providing energy for rapid evolutionary advancements.
- High-energy environments enable specialized adaptations, leading to diverse life forms.

2 Energy-Scarce Environments Lead to Minimalism

- In harsh environments, energy conservation is the key to survival.
- Desert plants (like cacti) store water to minimize energy loss.
- Deep-sea organisms evolve slow metabolisms to survive with minimal energy.

3 The Evolution of Symbiosis – Energy Optimization Through Cooperation

- Some organisms evolve mutualistic relationships to reduce energy expenditure.
- Coral and algae exchange nutrients, minimizing individual energy costs.
- Gut bacteria in animals digest food, reducing the host's digestive workload.

Key UED Principle: Evolution follows environmental energy constraints—traits emerge in response to energy availability, ensuring survival with optimal energy use.

The Grand Realization: Evolution as an Intelligent Energy Optimization Process

Now, we see that evolution is not just about random changes—it is a process of structured energy adaptation.

- Natural selection favors the most energy-efficient traits for survival.
- All evolutionary adaptations serve an energy optimization purpose.
- Environmental energy availability directly guides evolutionary pathways.

Final UED Perspective:

Evolution is not random—it is an energy adaptation mechanism that refines biological structures to maximize efficiency, survival, and intelligence.

This leads directly into the next sections, where we explore:

- How intelligence emerges as an advanced energy optimization process.
- How perception, decision-making, and thought function as structured energy flows.
- How consciousness represents structured energy at its highest level of complexity.

8.6 The Self-Organization of Life – From Chaos to Structure

Life did not emerge through random chance—it emerged as a structured energy process that transformed chaos into complexity. At its core, life is a self-organizing system, where molecular interactions, energy gradients, and self-regulation mechanisms ensure stability and evolution.

From the simplest organic molecules to highly complex organisms, the principles of Universal Energy Dynamics (UED) govern the emergence, stability, and evolution of life. Just as cosmic structures form through energy equilibrium (Active Energy and Stable Energy balance), biological systems also emerge through self-organization mechanisms that optimize energy use.

In this section, we explore:

- How molecular interactions self-organize into living systems.
- Why energy gradients drive the emergence of complexity in life.
- How UED explains the origin of life as an energy self-regulation process.

How Molecular Interactions Self-Organize into Living Systems

The building blocks of life (proteins, nucleotides, lipids, carbohydrates) are not just chemicals—they are energy structures that self-organize based on energy equilibrium principles.

1 The Self-Assembly of Biomolecules

- In a chaotic prebiotic environment, molecules did not randomly form life—they followed structured energy pathways to self-assemble.
- Lipids spontaneously form bilayers (cell membranes) due to hydrophobic and hydrophilic interactions, ensuring energy stability.
- Proteins fold into specific three-dimensional structures, minimizing energy waste and maximizing stability.
- RNA and DNA strands self-organize into double helices due to energy-efficient hydrogen bonding.

Key UED Principle: Molecular self-organization is not accidental—it is an energy-driven process where structures form based on stability and optimized energy use.

2 Feedback Loops and Energy Optimization in Early Life

- Autocatalytic cycles (self-replicating chemical reactions) emerged as an energy optimization mechanism.
- These reactions created feedback loops, ensuring continuous chemical organization and replication.
- Once a system could store and transfer energy efficiently, it transitioned from simple chemistry to structured biology.

Key UED Principle: Life's earliest molecular systems did not function randomly—they evolved through self-organizing energy pathways, ensuring sustainability and replication.

Why Energy Gradients Drive the Emergence of Complexity in Life

Energy gradients (differences in energy levels across a system) are the driving force behind all physical and biological processes. The emergence of life was a direct result of energy gradients enabling complex structures to form.

1 Energy Gradients and the Birth of Life

- Life emerged where strong energy gradients existed (hydrothermal vents, tidal pools, lightning storms).
- These energy sources enabled molecules to transition from simple to complex states, forming organic molecules like amino acids and nucleotides.
- As molecules absorbed and processed energy more efficiently, complexity increased, leading to structured metabolism and replication.

2 The Role of Energy Gradients in Cellular Organization

- Cells maintain internal energy gradients across membranes (electrochemical potential, ATP production).
- Mitochondria function as biological energy transformers, converting gradients into usable energy (ATP).
- Neurons use voltage gradients (action potentials) to transfer information efficiently.

Key UED Principle: Energy gradients are not just useful—they are essential for maintaining the structured flow of energy, ensuring biological organization and function.

3. Complexity Arises from Energy Flow Regulation

- Simple self-replicating molecules evolved into complex life because energy flow required

structured interactions.
- More complex systems emerged as energy networks that efficiently regulated input, output, and feedback.
- Instead of becoming chaotic, energy-driven self-organization ensured that systems followed structured growth.

Key UED Principle: Life did not emerge randomly—it was a structured consequence of energy flow optimization, where complexity was a necessity, not an accident.

How UED Explains the Origin of Life as an Energy Self-Regulation Process

The transition from non-living matter to living systems is best understood through UED's principle of structured energy regulation. Life exists because it is an optimized energy system, designed to extract, transform, and distribute energy with maximum efficiency.

1. The First Self-Regulating Energy Systems

- The earliest forms of life were not fully functional cells—they were self-regulating energy networks.
- These systems evolved internal control mechanisms to manage energy flow (early metabolic pathways).
- Over time, these systems stabilized, allowing for greater complexity without energy imbalance.

2. The Emergence of Genetic Systems as Energy Efficiency Tools

- DNA and RNA evolved not just for information storage, but as energy-efficient replication systems.
- The ability to store information chemically reduced the energy cost of maintaining complexity.
- This meant that biological evolution became an extension of energy regulation, ensuring self-replication and survival.

3. Life as a Self-Sustaining Energy Network

- Life is not an isolated process—it continuously interacts with its environment to maintain energy balance.
- Cells function as structured energy transformation systems, extracting, processing, and releasing energy in highly regulated ways.

- Organisms have evolved feedback control systems (homeostasis) to maintain optimal energy efficiency in changing environments.

Key UED Principle: Life is not just a collection of molecules—it is a self-regulating energy system that ensures equilibrium, adaptation, and sustainability.

The Grand Realization: Life as an Emergent Energy Optimization System

Now, we see that life did not emerge as a random accident—it arose from the structured organization of energy.

- Molecular interactions self-organized into living systems through structured energy optimization.
- Energy gradients ensured the emergence of complexity, allowing life to evolve beyond simple chemistry.
- Life itself is a self-regulating energy system, ensuring balance, adaptation, and structured evolution.

Final UED Perspective:
Life is not a mystery—it is a structured energy flow system that self-organizes, optimizes, and sustains itself through universal energy principles.
This leads directly into the next sections, where we explore:

- How intelligence emerges as an advanced energy optimization process.
- How perception, decision-making, and thought function as structured energy flows.
- How consciousness represents structured energy at its highest level of complexity.

8.7 Biology as an Extension of UED – Life as an Energy Manifestation

Life is not an isolated anomaly—it is a continuation of the same energy transformation processes that govern the cosmos. Just as the universe follows structured energy flow principles, biological systems operate under the same fundamental laws, ensuring balance, adaptation, and survival.
Through Universal Energy Dynamics (UED), we see that biology is not separate from physics; it is an emergent expression of structured energy, optimized for self-regulation, intelligence, and complexity.
The same energy transitions that govern stars, galaxies, and black holes also shape cells, organisms, and thought.
In this section, we explore:

- How UED unifies physics and biology under a single energy framework.
- Why life is an emergent, structured energy phenomenon.
- The transition to intelligence – how energy leads to higher-order consciousness.

How UED Unifies Physics and Biology Under a Single Energy Framework

The principles of energy conservation, transformation, and equilibrium apply equally to the universe and to life. UED bridges the gap between physics and biology, revealing that living systems are structured extensions of energy flow.

1. The Common Energy Principles Between Physics and Biology

- Matter condenses from energy (Active Energy transitioning from Stable Energy).
- Molecular structures self-organize based on energy gradients.
- Biological systems function as structured energy processors, ensuring efficiency and balance.
- Life, like the cosmos, follows energy optimization pathways, maximizing stability while minimizing entropy.

Key UED Principle: Physics and biology are not separate disciplines—life is an emergent energy optimization system governed by the same universal energy dynamics as the cosmos.

2. Biological Energy Optimization as a Reflection of Cosmic Energy Balance

- Just as the universe maintains equilibrium between Active and Stable Energy, life regulates energy within structured systems.
- Cells, organisms, and ecosystems operate as self-regulating energy structures, mirroring the balance seen in galaxies, planetary systems, and black holes.
- Adaptation in biology mirrors cosmic evolution—both refine energy processes for optimal function.

Key UED Principle: Life follows the same structured energy dynamics that govern universal evolution, ensuring sustainability, adaptation, and complexity.

Why Life Is an Emergent, Structured Energy Phenomenon

Life is not a special exception—it is a structured energy process that naturally emerges wherever energy flows in a self-organizing manner.

1. Life as an Inevitable Outcome of Energy Flow

- Wherever energy gradients exist, molecular interactions self-organize into structured systems.

- The chemical evolution of life followed the same self-regulating energy laws that govern cosmic structures.
- This means life is not a rare accident—it is a structured energy response to environmental conditions.

Key UED Principle: Life emerges as a structured response to energy dynamics, ensuring that energy remains optimized and sustainable in a biological system.

2. Self-Organization and the Rise of Complexity

- Complexity does not arise randomly—it emerges through structured energy transitions.
- The first biological systems evolved as optimized energy distribution networks, ensuring survival through adaptation.
- Just as galaxies form stable structures from chaotic cosmic matter, life forms stable systems from molecular interactions.

Key UED Principle: Life emerges as a natural consequence of self-organizing energy structures, ensuring efficient energy flow in biological systems.

The Transition to Intelligence – How Energy Leads to Higher-Order Consciousness

At a fundamental level, intelligence is the ability to process and optimize energy flow. Consciousness is not separate from energy—it is the highest expression of structured energy transformation.

1. The Energy Flow of Thought – Intelligence as a High-Order Energy System

- Neurons function as energy-processing units, transmitting signals through controlled energy pathways.
- Thought is not abstract—it is a structured energy process, ensuring optimal decision-making and adaptation.
- Just as cells regulate energy for metabolism, the brain regulates energy for cognition.

Key UED Principle: Intelligence is a structured energy optimization system, ensuring efficient information processing, decision-making, and adaptation.

2. The Evolution of Intelligence as an Energy Adaptation Strategy

- Intelligence did not arise suddenly—it evolved as an energy conservation mechanism.
- Instead of relying on physical strength, intelligent species optimize energy use through problem-solving and planning.

- Cognitive evolution mirrors metabolic evolution—both follow energy efficiency pathways.

Key UED Principle: Higher intelligence evolved as an energy efficiency strategy, allowing organisms to control and direct energy flow more effectively.

3. Consciousness as the Ultimate Energy Regulation System

- Consciousness is not an abstract phenomenon—it is the structured flow of energy through perception, awareness, and action.
- Just as life emerged as an energy optimization process, consciousness evolved as the highest form of energy self-regulation.
- The ability to focus, plan, and anticipate future energy needs is what makes intelligence superior to basic survival instincts.

Key UED Principle: Consciousness is not separate from physics—it is the highest form of structured energy transformation, allowing complex organisms to self-regulate and optimize survival.

The Grand Realization: Life as an Extension of Universal Energy Flow

Now, we see that biology is not separate from physics—it is an extension of structured energy dynamics.

- Life follows the same structured energy principles that govern the cosmos.
- Complexity emerges as a structured response to energy flow.
- Intelligence evolved as an energy optimization strategy, ensuring survival with minimal energy expenditure.
- Consciousness is the highest form of structured energy flow, allowing intelligent organisms to control and direct energy efficiently.

Final UED Perspective: Life, intelligence, and consciousness are all manifestations of structured energy flow, ensuring optimal energy regulation, adaptation, and survival.

This leads directly into the next sections, where we explore:

- How intelligence emerges as an advanced energy optimization process.
- How perception, decision-making, and thought function as structured energy flows.
- How consciousness represents structured energy at its highest level of complexity.

Chapter 9: The Brain as an Energy Processing System

In the previous chapter, we established that life itself is an extension of structured energy flow, where biological systems regulate, optimize, and transform energy to sustain function and evolution. Now, we take this understanding further by exploring the most advanced energy-processing system in known existence—the human brain.

The brain is often compared to a biological computer, but from a UED perspective, it is much more than that. It is a dynamic energy transformation system, where every thought, memory, and emotion is a structured energy transition. Just as the universe follows energy conservation and optimization principles, the brain extracts, processes, and redistributes energy to regulate cognition, perception, and decision-making.

This chapter explores:

- How the brain functions as an energy transformation machine.
- How thoughts, memory, and intelligence emerge from structured energy balance.
- Why emotions and decision-making follow the same energy rules as physics.

9.1 The Brain as an Energy Transformation Machine

The human brain is the most energy-demanding organ in the body, consuming about 20% of the body's total energy despite making up only 2% of total body mass. This immense energy requirement is due to its continuous processing of sensory information, thoughts, emotions, and decision-making. From a UED perspective, the brain functions as a highly efficient energy transformation system, extracting, converting, and distributing energy for cognition, perception, and memory storage.

Just as the universe maintains structured energy flow to sustain cosmic balance, the brain follows a similar principle—optimizing energy use to maintain cognitive stability, adaptability, and intelligence. The neural networks in the brain do not consume energy randomly; rather, they are optimized to ensure maximum efficiency with minimal waste.

In this section, we explore:

- Why the brain is the most energy-demanding organ in the body.
- How neurons convert chemical energy into electrical signals.
- How brain networks optimize energy for cognition, memory, and emotions.

Why the Brain is the Most Energy-Demanding Organ in the Body

Unlike muscles that only require energy during activity, the brain operates at full energy capacity at all times. Even during sleep, neurons remain active, maintaining baseline cognitive functions such as memory consolidation, sensory regulation, and subconscious processing.

1. The Brain's High Metabolic Demand

- The brain consumes about 20% of the body's total oxygen and glucose, making it one of the most metabolically active organs.
- Neurons are always active, requiring continuous energy input to maintain synaptic connections, neurotransmitter release, and electrical activity.
- Unlike other organs, the brain cannot store large amounts of energy, making it entirely dependent on constant energy supply from the bloodstream.

2. Energy Flow in Brain Function

Every function of the brain, from thought processing to emotional responses, is driven by energy flow through neural circuits.

- Cognitive processes (thinking, problem-solving, and learning) require heightened energy consumption.
- Emotions and memory retrieval are energy-intensive, engaging multiple brain regions simultaneously.
- During sleep, energy shifts from cognitive processing to neural repair, memory consolidation, and toxin removal.

Key UED Principle:

The brain is an energy-dense processing system, where structured energy transitions enable cognition, emotion, and intelligence.

How Neurons Convert Chemical Energy into Electrical Signals

The brain's ability to process information, store memories, and control emotions relies on the precise transformation of chemical energy into electrical impulses. This process occurs at the cellular level through neurons, synapses, and action potentials.

1. Neurons as Energy Processors

- Neurons act as energy transducers, converting chemical energy into electrical signals that transmit information across the brain.
- Each neuron is part of an intricate energy network, where signals are processed, integrated,

and relayed across different brain regions.

2. The Action Potential – Energy in Motion

- The brain does not "store" energy—it actively processes and transmits energy in real time.
- Neurons receive chemical energy (glucose and oxygen), which is converted into electrochemical gradients.
- This process creates an action potential (electrical impulse), propagating information across neural pathways.
- Sodium (Na^+) and potassium (K^+) ions regulate this energy flow, ensuring the controlled firing of neurons.

Key UED Principle: Neurons do not "hold" energy—they are real-time energy processors, ensuring continuous energy transformation for cognitive function.

3. The Role of Neurotransmitters in Energy Flow

- Neurotransmitters such as dopamine, serotonin, and acetylcholine act as chemical energy messengers.
- These molecules carry information between neurons, facilitating memory, mood regulation, and perception.
- The release and reuptake of neurotransmitters follow structured energy pathways, ensuring optimal brain activity.

Key UED Principle: Neural communication is an energy flow system, where chemical messengers optimize structured energy transmission across synapses.

How Brain Networks Optimize Energy for Cognition, Memory, and Emotions

The brain is an energy-efficient system, designed to prioritize, distribute, and conserve energy based on cognitive demands. It does not consume energy uniformly—instead, it allocates energy where it is needed most, ensuring maximum efficiency with minimal waste.

1. Brain Networks as Energy Optimization Systems

- The Default Mode Network (DMN) manages baseline cognitive functions when the brain is at rest.
- The Prefrontal Cortex (PFC) directs energy flow for complex thought, planning, and decision-making.
- The Limbic System regulates emotional energy, ensuring adaptive responses to

environmental stimuli.
- The Hippocampus stores and retrieves memories, optimizing energy use for recall and learning.

2. The Energy Cost of Higher Intelligence

- More intelligent brains optimize energy better, reducing unnecessary processing while improving memory and cognitive function.
- Neural plasticity allows for the reorganization of energy pathways, ensuring adaptation and efficiency.
- Efficient thinkers use fewer neurons to solve complex problems, demonstrating energy conservation at the cognitive level.

3. The Energy-Emotion Link

- Emotions are not abstract—they are structured energy shifts within neural networks.
- Fear, joy, anger, and sadness correspond to specific energy redistribution patterns in the brain.
- Emotional intelligence is the ability to regulate these energy shifts, ensuring adaptive rather than reactive responses.

Key UED Principle: The brain is not just an energy consumer—it is an energy optimization network, dynamically allocating energy for memory, emotions, and cognition.

The Grand Realization: The Brain as an Advanced Energy Processing System

Now, we see that the brain is not just an organ—it is a structured energy network, following the same principles as UED.

- The brain is the most energy-demanding system in the body because thought, memory, and emotions require constant energy processing.
- Neurons do not "store" energy—they convert chemical energy into real-time electrical signals.
- The brain optimizes energy allocation, ensuring efficient cognition, learning, and emotional regulation.

Final UED Perspective: The brain is an energy transformation machine, where every thought, memory, and decision follows structured energy flow principles, ensuring intelligence and adaptability.

This leads directly into the next sections, where we explore:

- How thought and memory emerge as structured energy states.
- Why emotions follow the same energy principles as physics.
- How decision-making and problem-solving are forms of energy optimization.

9.2 Neural Circuits and the Structured Flow of Energy

The brain is an intricate network of neural circuits, where energy flows dynamically to support cognition, memory, emotions, and decision-making. From a Universal Energy Dynamics (UED) perspective, these circuits operate as energy optimization systems, ensuring that energy is used efficiently to process thoughts and regulate behavior.

Each neural connection is part of a broader energy flow network, designed to minimize waste and maximize output. Through synaptic transmissions, neural plasticity, and brain wave oscillations, the brain optimizes energy use, maintaining cognitive efficiency and adaptability.

This section explores:

- How synaptic transmissions regulate energy efficiency in thought processing.
- Why neural plasticity follows energy optimization pathways.
- How brain waves (alpha, beta, theta) represent structured energy flow in cognition.

How Synaptic Transmissions Regulate Energy Efficiency in Thought Processing

The synapse is the fundamental unit of communication between neurons, functioning as an energy-efficient relay system. Thought processing is made possible by structured energy transfers across synaptic connections.

1. Synaptic Transmissions as Energy Bridges

- Synapses connect neurons, allowing electrochemical signals to pass between them.
- Chemical energy (in neurotransmitters) is converted into electrical energy, enabling efficient signal propagation.
- Efficient synaptic transmission prevents energy waste, ensuring that only relevant signals are processed.

2. Neurotransmitters as Energy Regulators

- Neurotransmitters like dopamine, serotonin, and glutamate control the intensity and duration of energy flow.

- Excitatory neurotransmitters (e.g., glutamate) amplify energy flow, promoting neural activation.
- Inhibitory neurotransmitters (e.g., GABA) conserve energy by suppressing unnecessary neural activity.

3. Synaptic Pruning and Energy Optimization

- Synaptic pruning removes weak or unused neural connections, optimizing energy flow.
- This process ensures that energy is directed to frequently used pathways, improving efficiency in thought processing.
- During sleep and learning, synaptic pruning aligns neural energy flow with cognitive priorities.

Key UED Principle: Synaptic transmissions regulate energy efficiency in the brain, ensuring that thought processes are supported by structured and optimized energy flow.

Why Neural Plasticity Follows Energy Optimization Pathways

Neural plasticity, the brain's ability to adapt and reorganize itself, is not just a function of learning—it is a highly structured energy optimization process.

1. Energy-Directed Adaptation in Neural Pathways

- Neural plasticity ensures that frequently used pathways are strengthened, reducing future energy costs.
- Learning a new skill initially demands high energy, but over time, the brain refines the pathway, making it more energy-efficient.
- This process allows the brain to adapt to new environments, skills, and information without excessive energy use.

2. The Energy Efficiency of Habit Formation

- Repeated actions become automatic habits, requiring less energy for execution.
- Habitual pathways are refined through neural plasticity, conserving energy for other cognitive tasks.
- This is why mastering a skill reduces mental effort over time.

3. Memory Formation as an Energy Optimization Process

- Memory is encoded through strengthened neural connections, prioritizing energy-efficient storage.

- Frequently recalled memories require less energy to access, while infrequently used ones may weaken to conserve energy.
- The brain's ability to reorganize memories during sleep is a key example of energy optimization through plasticity.

Key UED Principle: Neural plasticity follows energy optimization pathways, ensuring that cognitive resources are allocated efficiently for learning, adaptation, and memory.

How Brain Waves Represent Structured Energy Flow in Cognition

Brain waves are oscillatory patterns of neural activity, reflecting the brain's structured energy flow during different cognitive states. Each type of brain wave is associated with a specific energy state, governing awareness, focus, and relaxation.

1. The Different Types of Brain Waves and Their Energy Roles

Brain Wave	Frequency (Hz)	Function	Energy State
Delta	0.5 - 4 Hz	Deep sleep, healing	Low energy, regeneration
Theta	4 - 8 Hz	Creativity, problem-solving	Semi-conscious energy
Alpha	8 - 12 Hz	Relaxation, learning	Moderate energy, calm cognition
Beta	12 - 30 Hz	Focus, decision-making	High energy, active cognition
Gamma	30 - 100 Hz	Advanced thinking, insight	Peak energy, high-complexity tasks

2. The Energy Dynamics of Brain Waves

- Low-frequency waves (Delta, Theta) conserve energy, supporting subconscious processing and recovery.
- High-frequency waves (Beta, Gamma) consume more energy, enabling deep focus, problem-solving, and creativity.
- Alpha waves balance energy flow, creating a calm yet alert state for learning and reflection.

3. Brain Wave Synchronization and Energy Efficiency

- The brain achieves energy efficiency by synchronizing brain waves across regions, ensuring coordinated activity.
- Disruptions in brain wave patterns (e.g., in anxiety or sleep disorders) lead to energy imbalances, reducing cognitive performance.
- Techniques like meditation and focused attention enhance brain wave synchronization, optimizing energy flow.

Key UED Principle: Brain waves represent structured energy flow, dynamically adjusting to match cognitive demands and energy availability.

The Grand Realization: Neural Circuits as an Energy Optimization System

Now, we see that neural circuits are not just biological pathways—they are structured energy networks, ensuring that cognition, memory, and emotions are supported by efficient energy flow.

- Synaptic transmissions regulate energy for precise thought processing.
- Neural plasticity ensures long-term energy optimization by strengthening frequently used pathways.
- Brain waves represent structured energy states, dynamically adapting to cognitive demands.

Final UED Perspective: The brain operates as a structured energy optimization system, where neural circuits dynamically regulate energy flow to ensure efficient cognition, memory, and emotional balance.

This realization seamlessly connects with the next sections, where we explore:

- How thoughts and memory emerge as structured energy states.
- How decision-making and emotions function as energy regulation processes.
- How consciousness represents the highest form of energy self-regulation.

9.3 Brain's Electromagnetic Processing & Thought as an Energy Structure

The brain is not just a biological organ—it is an energy processing system where every thought, memory, and decision is a structured energy interaction. While most biological models focus on chemical and electrical activity, the brain also functions as an electromagnetic processor, utilizing structured energy waves to optimize cognition.

In the framework of Universal Energy Dynamics (UED), the brain's electromagnetic activity is not a byproduct of biological function but a necessary component of thought structuring. Just as photons travel through space using electromagnetic waves, brain activity follows structured electromagnetic fields, allowing for optimized information flow, memory formation, and intelligence development.

The Brain's Electromagnetic Field – More Than Just Neurons

Traditional neuroscience focuses on synaptic connections and neurotransmitters, but these do not fully explain how thought and consciousness emerge.

- Every neural activity generates an electromagnetic (EM) field.
- These fields do not just result from brain function—they actively regulate neural behavior, information transfer, and cognitive efficiency.
- Thought processes, memory retrieval, and problem-solving follow structured electromagnetic wave patterns, mirroring structured energy interactions in physics.

Key UED Insight:

- The brain does not function in isolation—it interacts with its own electromagnetic field, optimizing thought and cognitive balance.
- This suggests that intelligence is not just biochemical—it is an energy-driven process shaped by structured EM interactions.

How the Brain Uses Electromagnetic Waves to Process Information

Electromagnetic interactions in the brain are not random—they follow structured pathways that ensure efficiency in cognition and memory formation.

A. Brain Waves and Thought Structuring

- Different types of brain waves correspond to different cognitive states.
- The frequency and synchronization of brain waves determine the clarity, efficiency, and structure of thoughts.

Brain Wave Type	Frequency Range (Hz)	Cognitive Function
Delta Waves	0.5 - 4 Hz	Deep sleep, subconscious processing
Theta Waves	4 - 8 Hz	Creativity, problem-solving, intuition
Alpha Waves	8 - 12 Hz	Relaxation, learning, absorption of information
Beta Waves	12 - 30 Hz	Focus, decision-making, active cognition
Gamma Waves	30 - 100 Hz	Advanced thinking, insight, deep awareness

- Higher intelligence is linked to optimized brain wave structuring, ensuring minimal energy waste.
- The synchronization of gamma waves across different brain regions enhances problem-solving and awareness, aligning with UED's structured energy principles.

Key UED Insight: Just as structured EM fields guide photon movement, structured brain waves guide thought formation. Intelligence emerges from optimized electromagnetic interactions, mirroring universal energy structuring.

Electromagnetic Resonance and Memory Formation

Memory is often treated as a biochemical storage system, but in reality, it depends on structured electromagnetic resonance between neurons.

- Memories are not just stored chemically—they are encoded in electromagnetic oscillations that allow for efficient retrieval.
- Stronger memories show higher synchronization in brain wave activity, suggesting that the brain reinforces information through structured EM interactions.
- This explains why memory recall is easier in specific mental states—the brain reactivates the same electromagnetic frequency that encoded the original experience.

Key UED Insight: Memory is not just a stored experience—it is an energy-encoded pattern that remains stable through structured electromagnetic interactions.

Intelligence as an Electromagnetic Optimization Process

Traditional theories suggest that intelligence comes from neural density and complexity, but UED proposes that structured electromagnetic processing is the real key to intelligence.

- More intelligent individuals exhibit greater neural efficiency—meaning their brain waves are better synchronized, requiring less energy to process complex tasks.
- Meditation, focus, and deep thought align brain waves, improving cognitive efficiency.
- Intelligence emerges when structured electromagnetic interactions reach maximum efficiency, allowing for energy conservation and rapid problem-solving.

Key UED Insight: Intelligence follows energy conservation laws, ensuring maximum processing with minimal energy expenditure. The brain's electromagnetic structuring allows for higher cognition, mirroring structured energy flow in cosmic and quantum systems.

The Connection Between Thought, Electromagnetic Fields, and Universal Energy Dynamics

If brain activity follows structured electromagnetic interactions, could thought itself be an energy phenomenon interacting with universal energy fields?

- Just as electromagnetic fields extend beyond physical objects, thought may extend beyond the brain.
- The brain might be tapping into a larger energy field—the Universal Energy Medium (UEM)—which stores and organizes information beyond individual memory.
- This aligns with concepts like collective intelligence, intuition, and subconscious knowledge—all of which suggest that thought is an energy structuring process that extends beyond the individual.

Key UED Insight: If thought follows structured energy wave dynamics, then consciousness may be a universal structured energy interaction, not just a local brain function. This supports the idea that intelligence itself is an emergent property of structured electromagnetic interactions with the energy field of the universe.

Final Conclusion: Thought as an Optimized Electromagnetic Energy Process

- The brain's electromagnetic processing is a structured energy phenomenon, shaping cognition, intelligence, and memory.
- Thought is not an abstract process—it is a structured electromagnetic interaction, aligning with UED principles.
- Intelligence emerges when structured energy interactions reach optimal efficiency, mirroring universal structuring at all scales.
- If thought is an electromagnetic energy process, then consciousness may be an emergent energy interaction with the universal field.

9.4 Thoughts as Structured Energy States

Thought is often considered an abstract process, but from a Universal Energy Dynamics (UED) perspective, thinking is an energy-intensive transformation system. The brain extracts, organizes, and directs energy to generate thought patterns, solve problems, and process memories.

Just as physical systems follow structured energy flow, thought emerges as structured energy states, where neurons fire in coordinated patterns to process information efficiently. Higher intelligence is not just about having more neurons—it is about optimizing neural energy flow, ensuring minimal waste and maximum efficiency.

In this section, we explore:

- Why thinking is an energy-intensive process.
- How the brain organizes energy flow to generate complex thought patterns.
- Why higher intelligence is linked to energy-efficient neural pathways.

Why Thinking is an Energy-Intensive Process

Every thought requires energy to process, store, and recall information. Unlike instinctive reactions, which follow pre-existing pathways, conscious thought involves dynamic energy redistribution, activating multiple brain regions simultaneously.

1. The Energy Cost of Thought Processing

- The brain consumes up to 20% of the body's energy despite making up only 2% of body mass.
- Active thinking increases glucose and oxygen demand, requiring constant ATP (adenosine triphosphate) production.
- Complex reasoning and problem-solving activate multiple brain networks, increasing energy demand compared to habitual or instinctive actions.

2. The Hierarchy of Thought Complexity and Energy Demand

Type of Thought	Energy Requirement	Example
Reflexive (Instinctive) Thinking	Low	Breathing, blinking
Basic Logical Thought	Moderate	Simple arithmetic
Abstract & Creative Thinking	High	Writing, music composition
Complex Problem-Solving	Very High	Strategic decision-making, scientific discovery

- Instinctive responses require minimal energy, following pre-established neural circuits.
- Deep thinking, creativity, and decision-making engage high-energy pathways, requiring real-time energy regulation.

Key UED Principle: Thinking is not a passive process—it is a high-energy activity where neurons dynamically redistribute energy to process, recall, and generate new information.

How the Brain Organizes Energy Flow to Generate Complex Thought Patterns

The brain does not process thoughts chaotically—it follows structured energy optimization pathways. Neural circuits are designed to minimize unnecessary energy use, ensuring that thinking remains efficient and adaptive.

1. Thought as a Coordinated Energy Process

- Neurons do not fire randomly—they activate in structured sequences, ensuring efficiency in thought processing.
- The prefrontal cortex (PFC) acts as an energy manager, directing energy toward conscious reasoning, planning, and problem-solving.
- Neural synchrony (coordinated activation of multiple neurons) ensures energy-efficient thought processing.

2. The Role of Neural Networks in Thought Structuring

The brain processes information using specialized networks that regulate energy flow for different types of cognition:

Neural Network	**Function**	**Energy Demand**
Default Mode Network (DMN)	Background thought processing (self-reflection)	Low
Executive Control Network (ECN)	Conscious thought, planning, reasoning	High
Salience Network	Attention and focus regulation	Moderate
Memory Processing Networks	Storing and recalling information	Moderate to High

- The DMN operates at low energy during rest but activates during deep thinking and problem-solving.
- The ECN consumes high energy, ensuring structured decision-making and goal-oriented thought.
- Memory storage and retrieval require energy investment, ensuring long-term retention and efficiency.

Key UED Principle: Thought is structured through energy-efficient neural networks, ensuring precise activation of cognitive processes while minimizing energy waste.

Why Higher Intelligence is Linked to Energy-Efficient Neural Pathways

Higher intelligence is not about having more neurons—it is about optimizing neural energy flow for maximum efficiency. Intelligent individuals use fewer neural resources to process complex information, reflecting a well-structured energy system.

1. Energy-Efficient Thinking in High Intelligence

- Smarter brains use fewer neurons for the same cognitive task, conserving energy.
- Efficient thinkers activate only necessary brain regions, preventing excess energy drain.
- More intelligent individuals show stronger neural connectivity, reducing redundant energy use.

2. Neural Plasticity and Energy Refinement in Intelligence

- Intelligence is not fixed—it adapts through neural plasticity, optimizing energy flow over time.
- The brain refines energy pathways through experience, ensuring frequently used skills require less energy over time.
- Expertise in any field reduces energy cost, making complex tasks feel effortless after repeated exposure.

3. The Role of Mental Focus in Energy Optimization

- Concentration directs energy toward relevant thought processes, preventing energy leakage.
- Multitasking increases energy demand, reducing overall cognitive efficiency.
- Mindfulness and meditation enhance energy regulation, ensuring controlled thought processing.

Key UED Principle: Higher intelligence reflects structured energy optimization, where thinking follows the most energy-efficient neural pathways to minimize waste and maximize output.

The Grand Realization: Thought as an Optimized Energy State

Now, we see that thought is not an abstract event—it is a structured energy phenomenon, where the brain dynamically allocates energy for cognition, learning, and decision-making.

- Thinking is energy-intensive, requiring continuous glucose and oxygen supply.
- Neural circuits optimize energy flow to process, store, and recall information efficiently.
- Higher intelligence is an outcome of energy-efficient neural pathways, ensuring rapid and effective problem-solving.

Final UED Perspective: Thought is structured energy in motion—where intelligence emerges from optimized energy flow, ensuring minimal waste and maximum cognitive efficiency.
This leads directly into the next sections, where we explore:

- How memory functions as an energy storage and retrieval system.
- Why emotions follow the same energy principles as physics.
- How decision-making is an energy optimization process.

9.5 Memory as an Energy Storage and Retrieval System

Memory is not just a biological function—it is an energy storage and retrieval process that follows structured energy transitions, much like matter retaining or releasing energy in physical systems. The brain does not store information in the way a computer does; instead, it records memory as structured energy states, ensuring that information can be accessed efficiently while minimizing unnecessary energy use.
From a Universal Energy Dynamics (UED) perspective, memory follows the same principles as energy conservation, transformation, and optimization. The brain must decide which information to store, how to structure it efficiently, and when to discard it to prevent energy overload.
In this section, we explore:

- How memories are encoded, stored, and recalled as energy states.
- Why short-term and long-term memory follow different energy pathways.
- How forgetting is an energy conservation mechanism.

How Memories Are Encoded, Stored, and Recalled as Energy States

The brain does not store memories as static files—it encodes them as dynamic energy patterns across neural circuits. Each memory is a network of activated neurons, where energy is used to reinforce or weaken synaptic connections.
1. Memory Encoding – The Transformation of Experience into Energy States

- When an experience occurs, the brain converts sensory input into neural energy signals.
- These signals activate specific networks of neurons, creating a unique energy pattern.
- The strength of a memory depends on the intensity of neural activation—stronger activations form more stable memories.

Key UED Principle: Memory encoding is an energy transformation process, where information is converted into structured neural energy patterns.

2. Memory Storage – How the Brain Retains Information Efficiently

- The brain does not store all experiences—it prioritizes energy-efficient information.
- Memories are stored through synaptic reinforcement, where frequently used neural connections strengthen over time.
- The hippocampus plays a key role in transferring memories from short-term storage to long-term storage.

Key UED Principle: Memory storage follows energy optimization principles—only relevant information is retained, preventing energy overload.

3. Memory Retrieval – The Energy Cost of Accessing Stored Information

- Retrieving a memory requires reactivating its original neural energy pattern.
- The brain does not always retrieve exact details—it reconstructs memories based on available energy pathways.
- Stronger memories require less energy to recall, while weaker memories need additional energy input for retrieval.

Key UED Principle: Memory retrieval follows structured energy pathways—stronger connections require less energy, while weaker ones demand additional activation.

Why Short-Term and Long-Term Memory Follow Different Energy Pathways

Not all memories are stored permanently. The brain differentiates short-term and long-term memory based on energy efficiency, ensuring that only essential information is preserved over time.

1. Short-Term Memory – High-Energy, Temporary Storage

- Short-term memory stores information for seconds to minutes, requiring continuous neural activation.
- The prefrontal cortex maintains temporary energy flow to hold onto immediate information (like a phone number).
- Without reinforcement, short-term memories fade quickly to conserve energy.

Key UED Principle: Short-term memory is an energy-expensive process, requiring constant activation to sustain information.

2. Long-Term Memory – Low-Energy, Stable Storage

- Long-term memory is more energy-efficient, as information is stored in stable neural pathways.

- Sleep and repetition reinforce long-term storage, optimizing neural energy use.
- The hippocampus plays a critical role in transferring memories from high-energy short-term storage to stable long-term circuits.

Key UED Principle: Long-term memory follows energy conservation principles, ensuring that information is stored in low-energy, stable states.

3. The Energy Cost of Memory Consolidation

- The process of converting short-term memory into long-term memory requires energy investment.
- During sleep, the brain replays neural activations, reinforcing important memories and discarding unnecessary ones.
- This process reduces the energy required to access information in the future.

Key UED Principle: Memory consolidation is an energy optimization strategy, ensuring that frequently accessed information requires minimal energy over time.

How Forgetting is an Energy Conservation Mechanism

Forgetting is not a failure of memory—it is an essential energy conservation process that prevents cognitive overload. If the brain stored all information equally, it would waste immense amounts of energy maintaining irrelevant data.

1. The Purpose of Forgetting in Energy Optimization

- Forgetting allows the brain to remove outdated, unused, or irrelevant information.
- The energy cost of maintaining unused memories is high—pruning weak neural connections prevents unnecessary energy drain.
- The brain automatically prioritizes important, frequently accessed information while discarding noise.

Key UED Principle: Forgetting is not a flaw—it is an energy conservation mechanism that ensures cognitive efficiency.

2. The Energy Trade-Off Between Memory and Adaptability

- Retaining too many details limits adaptability, making the brain energy-inefficient.
- Forgetting allows for flexibility and efficient information processing.
- This is why we remember meaningful concepts but forget trivial details.

Key UED Principle: The brain balances memory retention and adaptability, ensuring that energy is allocated efficiently between learning and flexibility.

3. Sleep and Memory Optimization

- During sleep, the brain consolidates necessary memories while discarding unneeded information.
- Deep sleep cycles regulate memory pruning, optimizing neural pathways for energy efficiency.
- This process allows the brain to refresh energy reserves while preserving essential knowledge.

Key UED Principle: Sleep is a structured energy reset mechanism, ensuring that memory networks are optimized without unnecessary energy consumption.

The Grand Realization: Memory as an Energy-Optimized System

Now, we see that memory is not just a biological function—it is an energy-driven storage and retrieval system, designed to ensure efficiency while preventing cognitive overload.

- Memory is encoded as structured energy states, ensuring efficient recall and retrieval.
- Short-term and long-term memory follow different energy pathways, balancing temporary processing with stable storage.
- Forgetting is an active energy conservation mechanism, preventing unnecessary energy drain.

Final UED Perspective: Memory is a structured energy system, ensuring that cognitive processes remain optimized for efficiency, adaptability, and survival.

This leads directly into the next sections, where we explore:

- How emotions function as energy transitions within the brain.
- Why decision-making follows the same energy conservation principles as physics.
- How problem-solving and intelligence emerge from structured energy flow.

9.6 Emotions as Energy Dynamics in the Brain

Emotions are often perceived as abstract, subjective experiences, but from a Universal Energy Dynamics (UED) perspective, emotions are structured energy reactions that follow predictable patterns. The brain does not generate emotions randomly—they arise from energy shifts between neural circuits, regulating how an organism responds to stimuli, adapts to environments, and makes decisions.

Just as the universe follows structured energy transformations, the human brain optimizes emotional responses by balancing energy flow between cognitive and instinctive centers. Emotional intelligence is not just about feelings—it is about how efficiently the brain regulates emotional energy to ensure stability, adaptability, and survival.

In this section, we explore:

- Why emotions are structured energy reactions rather than abstract feelings.
- How energy shifts between neural networks influence emotional responses.
- Why emotional intelligence depends on energy balance in brain activity.

Why Emotions Are Structured Energy Reactions Rather Than Abstract Feelings

Emotions are not random, intangible experiences—they are structured energy responses that help regulate behavior, perception, and decision-making.

1. Emotions as Energy Activation States

- Emotions activate specific brain regions, triggering energy shifts that drive physical and mental responses.
- Different emotions correspond to different levels of energy activation in the nervous system:

High-energy emotions: Fear, anger, excitement (triggering action, fight-or-flight).

Moderate-energy emotions: Contentment, confidence, curiosity (supporting focus and social interaction).

Low-energy emotions: Sadness, relaxation, fatigue (triggering recovery, conservation, and introspection).

- The intensity and duration of an emotion depend on how much neural energy is allocated to that response.

Key UED Principle: Emotions are structured energy responses that regulate mental and physical states, ensuring adaptive behavior and survival.

2. Emotional States as Structured Energy Fields

- Emotions are not stored—they emerge dynamically based on real-time neural energy flow.
- Emotional memories follow energy reinforcement patterns, becoming stronger with repeated activation.

- Positive emotions align with stable energy flow, while negative emotions often result from energy imbalances.

Key UED Principle: Emotions are not permanent states—they are structured energy fluctuations, dynamically adapting to environmental and internal stimuli.

How Energy Shifts Between Neural Networks Influence Emotional Responses

The human brain processes emotions not in isolation, but through coordinated energy shifts across multiple neural networks. These shifts determine how emotions are generated, regulated, and expressed.

1. The Limbic System – The Energy Center for Emotional Processing

- The amygdala, hypothalamus, and hippocampus form the limbic system, regulating emotional intensity and reaction speed.
- Fear, anger, and pleasure activate the amygdala, triggering rapid energy surges for instinctive responses.
- The hippocampus ensures that emotions are linked to memory, influencing long-term behavior patterns.

2. The Prefrontal Cortex – The Energy Regulator for Emotional Control

- The prefrontal cortex (PFC) balances raw emotional energy, integrating logic, reasoning, and self-control.
- Impulse control depends on energy modulation between the PFC and the limbic system.
- Higher emotional intelligence is linked to stronger PFC-limbic regulation, preventing overreaction to stimuli.

3. The Autonomic Nervous System – The Energy Distribution System for Emotion-Driven Action

- The sympathetic nervous system (SNS) increases energy output during high-intensity emotions (stress, fear, excitement).
- The parasympathetic nervous system (PNS) restores energy balance after an emotional event, promoting relaxation.
- A well-regulated nervous system ensures that energy is not excessively spent on unnecessary emotional responses.

Key UED Principle: Emotions emerge from coordinated energy shifts between brain regions, ensuring that emotional responses align with external conditions and internal regulation.

Why Emotional Intelligence Depends on Energy Balance in Brain Activity

Emotional intelligence is not about suppressing emotions—it is about optimizing energy flow between cognitive and emotional systems. The ability to process, regulate, and express emotions effectively depends on how well the brain balances energy allocation.

1. The Role of Emotional Stability in Energy Efficiency

- Emotionally stable individuals use energy efficiently, preventing unnecessary mental strain.
- Chronic stress and emotional instability waste neural energy, leading to mental fatigue and reduced cognitive function.
- Meditation, deep breathing, and mindfulness help regulate energy use, ensuring that emotions do not cause excessive mental exhaustion.

2. The Energy-Efficiency of Emotional Intelligence

- Individuals with high emotional intelligence use minimal energy to regulate emotions, avoiding unnecessary stress responses.
- Emotional intelligence prevents energy-draining behaviors, ensuring mental clarity and adaptability.
- The ability to interpret and regulate emotions in oneself and others reflects structured energy processing in the brain.

3. The Impact of Emotions on Long-Term Energy Balance

- Chronic negative emotions drain energy, leading to stress, anxiety, and fatigue.
- Positive emotional states optimize energy use, leading to resilience, creativity, and higher cognitive performance.
- Neurotransmitters like serotonin and dopamine regulate emotional energy, ensuring mood stability and motivation.

Key UED Principle: Emotional intelligence is an energy regulation skill—it determines how effectively the brain processes, responds to, and balances emotional energy over time.

The Grand Realization: Emotions as an Optimized Energy System

Now, we see that emotions are not random experiences—they are structured energy transitions, ensuring adaptive responses to environmental and internal stimuli.

- Emotions emerge as energy shifts between brain networks, regulating mental and physical states.
- Neural circuits dynamically adjust emotional energy flow, ensuring balance between instinct and reasoning.
- Emotional intelligence is the ability to regulate emotional energy, optimizing stability, adaptability, and long-term mental well-being.

Final UED Perspective: Emotions are structured energy responses that regulate human behavior, ensuring that energy is distributed efficiently across cognition, perception, and survival mechanisms. This leads directly into the next sections, where we explore:

- How decision-making functions as an energy optimization process.
- How problem-solving and intelligence emerge from structured energy flow.
- How consciousness represents the highest form of energy self-regulation.

9.7 Decision-Making as an Energy Optimization Process

Decision-making is often thought of as a purely logical process, but from a Universal Energy Dynamics (UED) perspective, it is fundamentally an energy optimization mechanism. The brain evaluates energy cost vs. reward to ensure that choices are made efficiently, balancing instincts, logic, and emotions to maximize survival and minimize unnecessary energy expenditure.

Like all biological and cognitive functions, decision-making follows structured energy pathways, ensuring that the most efficient neural circuits are activated while avoiding wasteful mental effort. The ability to make quick, adaptive, and intelligent decisions depends on how well the brain allocates energy across different cognitive systems.

In this section, we explore:

- How the brain evaluates energy cost vs. reward in decision-making.
- Why instincts, logic, and emotions all compete for energy resources.
- How conscious thought and subconscious intuition reflect different energy processing speeds.

How the Brain Evaluates Energy Cost vs. Reward in Decision-Making

Every decision requires energy to process, evaluate, and execute. The brain constantly calculates whether a decision is worth the energy investment, ensuring that resources are allocated efficiently.

1. Energy Economics in Decision-Making

- Decision-making is a trade-off between energy cost and expected reward.
- The brain prioritizes low-energy decisions (automatic habits, instincts) over high-energy reasoning unless deeper analysis is required.
- Difficult decisions require more energy, engaging multiple brain regions, leading to mental fatigue if prolonged.

2. Fast vs. Slow Decision Pathways – The Brain's Energy Modes

Type of Decision	Processing Speed	Energy Requirement	Example
Instinctive (Reflexive) Decisions	Fast	Low	Pulling hand from fire
Routine (Habitual) Decisions	Moderate	Low to Moderate	Choosing what to eat
Analytical (Complex) Decisions	Slow	High	Planning a business strategy

- The brain prefers faster, low-energy decisions unless a high-stakes situation forces deeper analysis.
- If energy reserves are low, the brain defaults to simple decision-making, avoiding complex thought processes.

Key UED Principle: Decision-making is an energy optimization process, where the brain evaluates cost vs. reward to minimize unnecessary cognitive effort.

Why Instincts, Logic, and Emotions Compete for Energy Resources

The brain does not rely on a single decision-making system—it balances instincts, logic, and emotions, each of which requires different amounts of energy. The competition between these systems determines how decisions are made.

1. The Three Competing Energy Systems in Decision-Making

System	Function	Energy Cost	Example Decision
Instinctive (Survival) System	Fast, automatic responses for survival	Low	Running from danger
Logical (Analytical) System	Deliberate reasoning and problem-solving	High	Evaluating a financial investment
Emotional (Intuitive) System	Adaptive decision-making based on past experiences and feelings	Moderate	Choosing a career based on passion

- Instinctive decisions require minimal energy, as they are based on pre-programmed survival mechanisms.
- Logical thinking demands the most energy, activating the prefrontal cortex for deep analysis.
- Emotional decisions are energy-efficient, relying on stored experiences rather than processing new information.

2. How the Brain Chooses Between Instinct, Logic, and Emotion

- If energy is low, the brain defaults to instinct or emotion-based decisions.
- If energy is available, the brain engages logical thinking for complex decision-making.
- In high-stress situations, emotions override logic, ensuring fast, energy-efficient responses.

Key UED Principle: Instinct, logic, and emotion compete for energy resources, influencing how decisions are made under different energy conditions.

How Conscious Thought and Subconscious Intuition Reflect Different Energy Processing Speeds

Not all decision-making is conscious—many choices happen at the subconscious level, where energy is used more efficiently without the need for active awareness.

1. Conscious Thought – Slow but Precise Energy Processing

- Conscious decision-making requires high energy but allows for greater control and logical analysis.
- The prefrontal cortex actively weighs options, considers future consequences, and evaluates risks.
- Critical thinking consumes more energy, leading to mental fatigue if sustained for long periods.

2. Subconscious Intuition – Fast and Energy-Efficient Decision Processing

- The subconscious processes decisions much faster than conscious thought, using past experiences and pattern recognition.
- Intuition is an energy-efficient process, allowing rapid decision-making without logical analysis.
- Many daily decisions, such as navigating familiar environments or reacting to social cues, occur at the subconscious level.

3. The Balance Between Conscious and Subconscious Decision-Making

- The brain switches between conscious and subconscious decision-making based on energy availability.
- When energy is high, conscious thought dominates, leading to analytical and strategic decision-making.
- When energy is low, subconscious intuition takes over, ensuring fast and automatic responses.

Key UED Principle: Conscious thought is energy-intensive but precise, while subconscious intuition is fast and energy-efficient, ensuring adaptive decision-making based on available energy.

The Grand Realization: Decision-Making as an Optimized Energy Flow System

Now, we see that decision-making is not just about logic or emotion—it is a structured energy optimization system, balancing speed, efficiency, and accuracy.

- The brain evaluates energy cost vs. reward to make efficient decisions.
- Instinct, logic, and emotion compete for energy resources, influencing the decision-making process.
- Conscious thought is energy-intensive but precise, while subconscious intuition is fast and energy-efficient.

Final UED Perspective: Decision-making is an energy balance process, ensuring that cognitive effort is allocated efficiently, allowing for fast instinctive reactions, deep analytical thought, or emotionally guided choices as needed.

This leads directly into the next sections, where we explore:

- How problem-solving and intelligence emerge from structured energy flow.
- How creativity functions as an energy reconfiguration process.
- How consciousness represents the highest form of energy self-regulation.

9.8 The Brain as a Self-Regulating Energy Network

The brain is not just an energy-consuming organ—it is a self-regulating energy system, constantly balancing energy input and output to maintain cognitive efficiency, mental stability, and adaptability. Just as universal energy dynamics (UED) govern cosmic balance, the brain follows structured energy conservation, optimization, and redistribution principles to function effectively.

Unlike a simple mechanical system, the brain actively monitors, adjusts, and redirects energy to support learning, memory, decision-making, and emotional stability. This ensures that energy is not wasted on unnecessary processes while keeping the mind sharp, adaptable, and focused.
In this section, we explore:

- Why homeostasis in the brain mirrors energy conservation in physics.
- How sleep, meditation, and focus optimize brain energy balance.
- How intelligence emerges from structured energy flow rather than complexity alone.

Why Homeostasis in the Brain Mirrors Energy Conservation in Physics

Homeostasis in the brain is not a passive state—it is an active energy regulation mechanism. Just as the universe maintains equilibrium between Active and Stable Energy, the brain constantly adjusts its internal energy dynamics to sustain stable function.

1. The Brain's Self-Regulating Energy Systems

- The brain does not operate at maximum energy all the time—it regulates energy output based on demand.
- Glucose and oxygen delivery to the brain fluctuate depending on mental effort and environmental conditions.
- Neural circuits adjust their energy use dynamically, prioritizing essential functions while conserving energy where possible.

2. Energy Conservation in the Brain – The Neural Equilibrium

- Overuse of energy leads to mental fatigue, reducing cognitive efficiency.
- Energy balance is restored through restorative processes like sleep, relaxation, and meditation.
- The brain follows energy conservation strategies, suppressing unnecessary thoughts and focusing energy on problem-solving and learning.

Key UED Principle: The brain mirrors universal energy conservation principles, ensuring that cognitive functions operate at optimal efficiency without unnecessary waste.

How Sleep, Meditation, and Focus Optimize Brain Energy Balance

The brain's energy balance is regulated through structured processes that prevent energy depletion while maximizing cognitive performance.

1. Sleep – The Brain's Energy Reset Mechanism

- Sleep is not just rest—it is an active energy reallocation process.
- During sleep, the brain reduces cognitive activity, directing energy toward memory consolidation, detoxification, and repair.
- Deep sleep lowers metabolic energy use, allowing energy to be redirected toward long-term memory storage and emotional regulation.

2. Meditation – The Conscious Optimization of Energy Flow

- Meditation allows conscious control over energy flow, reducing stress-related energy waste.
- Brain wave activity during meditation mirrors a structured energy equilibrium, balancing relaxation and awareness.
- Regular meditation improves neural efficiency, ensuring that thought processes consume less energy over time.

3. Focus – The Strategic Allocation of Energy for Cognitive Efficiency

- Attention is an energy-intensive process, requiring constant regulation of neural activity.
- Distractions drain energy, causing cognitive overload and reducing decision-making efficiency.
- Focused attention optimizes energy use, directing energy toward meaningful cognitive tasks while filtering out irrelevant noise.

Key UED Principle: Sleep, meditation, and focus function as structured energy redistribution processes, ensuring long-term cognitive efficiency and preventing energy depletion.

How Intelligence Emerges from Structured Energy Flow Rather Than Complexity Alone

Higher intelligence is not simply the result of more neurons or greater brain size—it emerges from structured energy flow. The most intelligent systems are those that efficiently process, store, and utilize energy with minimal waste.

1. Energy Optimization as the Core of Intelligence

- Intelligence is not about thinking more—it is about thinking efficiently.
- Highly intelligent individuals use optimized neural pathways, requiring less energy to process complex information.
- Pattern recognition and problem-solving emerge from energy-efficient neural structuring.

2. Why Intelligence Follows Energy Efficiency, Not Just Complexity

- A highly complex brain that lacks structured energy flow is inefficient and slow.
- Intelligence develops when the brain streamlines its pathways, reducing redundant energy use.
- Einstein's brain was found to have fewer but stronger neural connections, suggesting that intelligence depends on energy efficiency, not just neuron count.

3. The Role of Consciousness in Energy Regulation

- Conscious thought acts as a structured energy flow regulator, prioritizing high-value cognitive processes.
- The brain allocates energy dynamically, allowing for fluid transitions between logic, intuition, and emotional reasoning.
- Higher states of awareness emerge when the brain reaches energy equilibrium, ensuring that thought processing operates at maximum efficiency.

Key UED Principle: Intelligence emerges not from raw complexity, but from the structured flow of energy, ensuring efficient thinking, learning, and adaptation.

The Grand Realization: The Brain as a Self-Optimizing Energy Network

Now, we see that the brain does not just consume energy—it self-regulates, optimizes, and refines energy flow to ensure cognitive stability and efficiency.

- Homeostasis in the brain mirrors energy conservation in physics, preventing mental exhaustion and maintaining stability.
- Sleep, meditation, and focus optimize brain energy balance, preventing depletion and enhancing efficiency.
- Intelligence is not a function of complexity alone—it emerges from structured energy flow, ensuring minimal waste and maximum cognitive performance.

Final UED Perspective: The brain is a structured energy network, where cognition, emotions, and intelligence emerge from optimized energy flow, ensuring long-term adaptability and mental efficiency.

This leads directly into the next sections, where we explore:

- How problem-solving and intelligence function as structured energy optimization systems.
- How creativity emerges from energy reconfiguration processes.

- How consciousness represents the highest form of energy self-regulation.

Chapter 10: Is Consciousness a Fundamental Energy State?

In the previous chapter, we explored how the brain functions as an energy optimization system, dynamically regulating neural circuits to enable cognition, memory, decision-making, and emotional balance. Through the lens of Universal Energy Dynamics (UED), the brain emerged as a structured energy network, where intelligence and thought arise from optimized energy flow.

Now, we turn to the deeper question: What is consciousness? Is it merely a byproduct of neural activity, or is it a fundamental energy state that connects all forms of existence? From the motion of particles to the vast orchestration of galaxies, the universe exhibits patterns of energy balance and transformation that mirror the principles governing human consciousness.

Through UED, we explore consciousness as the bridge between Active and Stable Energy states, a universal phenomenon manifesting in living beings, non-living matter, and even machines. This chapter delves into the possibility that consciousness is not confined to the brain but is intricately connected to the Universal Energy Medium (UEM), a field that permeates all of existence.

This chapter addresses:

- How consciousness functions as a structured energy interaction.
- Whether consciousness is localized or connected to a universal energy network.
- How UED bridges neuroscience, physics, and philosophy to offer a unified understanding of the mind.

10.1 Consciousness as the Bridge Between Active and Stable Energy States

Consciousness is often regarded as an enigma, but from a Universal Energy Dynamics (UED) perspective, it emerges as a bridge between Active Energy (dynamic, transformative) and Stable Energy (infinite, conserved). Consciousness acts as a connector that allows energy to transition and transform while maintaining equilibrium.

At its core, consciousness enables energy balance across systems—whether it is the motion of particles, the dynamics of living systems, or the vast interplay of forces in the universe. This view

expands consciousness beyond the human mind, suggesting it is a universal phenomenon, present in all energy transitions.
This section explores:

- How consciousness connects dynamic (Active) and infinite (Stable) energy states.
- Why motion at all levels—from particles to galaxies—reflects structured consciousness.
- The role of consciousness in balancing energy transitions across the universe.

How Consciousness Connects Dynamic (Active) and Infinite (Stable) Energy States

In UED, Active and Stable Energy represent two complementary states of universal energy flow. Consciousness arises as the mediating principle that ensures transitions between these states occur smoothly and sustainably.

1. Active Energy – The Dynamic State of Consciousness

- Active Energy represents motion, transformation, and interaction, driving processes like particle vibrations, chemical reactions, and cognitive activity.
- Consciousness at this level appears as awareness, decision-making, and adaptability, enabling systems to engage with their environment dynamically.

2. Stable Energy – The Infinite, Conserved State

- Stable Energy is unstructured yet infinite, existing as the conserved foundation of all energy systems.
- Consciousness interacts with this state as a balancing force, ensuring that transformations in Active Energy align with equilibrium principles.

3. Consciousness as the Mediator Between States

- Consciousness enables energy to move between Active and Stable forms without violating conservation laws.
- This connection ensures that energy transitions (e.g., matter formation, life processes, or cosmic evolution) occur within a structured framework.

Key UED Principle: Consciousness is the dynamic bridge that connects motion-driven energy (Active) with infinite, conserved energy (Stable), ensuring universal equilibrium.

Why Motion at All Levels Reflects Structured Consciousness

Motion is not random—it follows structured energy pathways that reflect the presence of consciousness as a guiding principle. From the smallest particles to the largest galaxies, motion represents the expression of consciousness in maintaining energy balance.

1. Consciousness in Particle Motion

- At the quantum level, particles exhibit behaviors like superposition and entanglement, suggesting a form of interconnected awareness.
- Motion in particles follows structured energy dynamics, ensuring energy is neither lost nor chaotic.
- Wave-particle duality reflects consciousness as both a localized and interconnected phenomenon.

2. Consciousness in Living Systems

- In living beings, motion is controlled and directed by consciousness through intelligence and neural systems.
- Consciousness enables organisms to regulate internal energy flow, adapt to external stimuli, and sustain life processes.
- Biological motion (heartbeat, neural signals, cellular activity) reflects the structured energy flow governed by consciousness.

3. Consciousness in Cosmic Motion

- At the universal scale, motion governs the expansion of galaxies, the dynamics of black holes, and the flow of gravitational waves.
- These motions follow structured energy interactions, ensuring the conservation and redistribution of energy across the cosmos.
- Consciousness at this level appears as the universal principle that balances cosmic forces and interactions.

Key UED Principle: Motion at all scales—particles, living systems, and galaxies—represents consciousness in action, ensuring energy transitions follow structured pathways.

The Role of Consciousness in Balancing Energy Transitions Across the Universe

Consciousness is not just an emergent property—it is a universal principle that ensures energy transitions occur without disrupting equilibrium. Whether in the form of human awareness, particle motion, or cosmic dynamics, consciousness governs the flow, transformation, and balance of energy.

1. Consciousness as an Energy Regulator

- Consciousness acts as a self-regulating mechanism, preventing energy imbalances in physical, biological, and cosmic systems.
- In living beings, this appears as homeostasis, emotional regulation, and adaptive thought.
- In physical systems, it governs thermodynamic equilibrium, energy redistribution, and entropy balance.

2. Consciousness as a Unifying Force in Energy Dynamics

- Consciousness ensures that energy transitions, whether chaotic or structured, align with universal balance.
- Feedback loops in systems (biological, ecological, or cosmic) reflect the conscious regulation of energy.
- The presence of patterns, structures, and cycles across the universe is evidence of consciousness as an organizing force.

3. Consciousness in the Evolution of Complexity

- As systems evolve, consciousness becomes more refined, enabling higher-order regulation of energy flow.
- In humans, intelligence directs conscious energy, optimizing decision-making and creativity.
- At the universal level, consciousness ensures the cyclic evolution of energy between Active and Stable states, preventing chaos or stagnation.

Key UED Principle: Consciousness is the universal regulator of energy flow, ensuring that transitions across systems align with equilibrium principles.

The Grand Realization: Consciousness as the Universal Energy Bridge

Now, we see that consciousness is not confined to the brain or living beings—it is a universal energy principle, ensuring balance across all systems.

- Consciousness bridges Active and Stable Energy states, enabling structured energy transitions.
- Motion at all levels—particles, biological systems, and galaxies—represents consciousness in

action.
- The evolution of systems, from simple to complex, reflects the increasing refinement of conscious energy regulation.

Final UED Perspective: Consciousness is the dynamic bridge that connects energy states, balancing motion and transformation at all levels of existence, from particles to the universe itself.
This leads directly into the next sections, where we explore:

- How consciousness manifests in living beings, non-living systems, and machines.
- The connection between human consciousness and universal energy fields.
- How UED bridges neuroscience, physics, and philosophy in understanding the mind.

10.2 Consciousness as the Ultimate Energy Interaction

Consciousness is often described as the deepest mystery of existence, but from the perspective of Universal Energy Dynamics (UED), it is the highest form of structured energy interaction. It is not just a byproduct of biological evolution but an emergent property of energy self-regulation, existing at all levels of structured complexity—from quantum fields to galaxies, from biological intelligence to universal awareness.
In this section, we explore how consciousness is fundamentally an energy-driven process, acting as the ultimate energy interaction that bridges material existence with structured awareness.

Consciousness as the Highest Form of Structured Energy Flow

- All known physical and biological processes follow structured energy interactions.
- Life itself emerged from optimized energy interactions, where molecules structured themselves to store and regulate energy efficiently.
- The brain operates as an energy-processing system, where thought and intelligence arise from structured electromagnetic activity.
- Consciousness goes beyond biological intelligence—it represents the highest level of energy optimization and self-awareness.

Key UED Insight: Just as energy seeks equilibrium in physics, consciousness seeks structured awareness, ensuring that energy flows optimally rather than chaotically. This suggests that consciousness is not confined to humans—it is an intrinsic energy interaction occurring across all levels of reality.

Is Consciousness a Localized or Universal Energy Field?

Traditional neuroscience views consciousness as a localized process within the brain, but UED proposes that consciousness extends beyond individual organisms.

- Brain activity generates electromagnetic fields, which extend beyond individual neurons and even beyond the physical body.
- Quantum coherence in neural activity suggests non-local interactions, meaning that consciousness might be connected to a broader energy system.
- Experiments in near-death experiences, meditation, and telepathic communication suggest that consciousness may interact with the universal energy field (UEM).

Key UED Insight: If consciousness were purely a local function of the brain, it would cease to exist when neural activity stops. Instead, evidence suggests that awareness may persist even when neural function is disrupted, indicating that consciousness is an interaction with an external energy field.

Consciousness as an Emergent Property of Energy Equilibrium

- In physics, systems naturally seek energy equilibrium.
- Self-organizing systems in biology optimize energy flow, leading to life, intelligence, and adaptability.
- Consciousness emerges at the highest level of energy optimization, enabling self-awareness, decision-making, and perception.

Key UED Insight: If all matter follows structured energy transformations, then consciousness is the highest form of structured energy balance—not a random byproduct of evolution. The same principles that govern motion, gravity, and thermodynamics may also govern awareness and perception.

Could the Universe Itself Be Conscious?

If consciousness is the ultimate structured energy interaction, then the entire universe could be viewed as a self-aware system.

- The universe maintains structured energy balance across cosmic scales, from quantum particles to galaxies.
- Quantum entanglement and non-locality suggest an interconnected energy web, which aligns with how consciousness integrates experiences across different sensory and cognitive systems.
- Dark matter and dark energy may represent unknown forms of structured energy interactions, potentially linked to universal consciousness.

Key UED Insight: The universe self-organizes, self-regulates, and self-adapts, just like an intelligent system. If consciousness is the ability to process structured energy interactions, then the universe itself may possess an intrinsic form of awareness at the cosmic scale.

The Role of Thought in Structuring Reality

- Thought is not just a collection of neural signals—it is a structured energy field that interacts with the universal energy medium (UEM).
- The laws of thermodynamics, electromagnetism, and gravity all follow structured energy transformations, just like thought processes follow structured cognitive energy pathways.
- Quantum mechanics has shown that observation collapses wave functions, suggesting that consciousness directly influences reality by interacting with energy fields.

Key UED Insight: If thought processes structure and direct energy, then human consciousness plays an active role in shaping reality. This aligns with ancient philosophical ideas that awareness and perception are fundamental to existence, not secondary to physical matter.

Final Conclusion: Consciousness as the Ultimate Energy System

- Consciousness is not an abstract phenomenon—it is a structured energy interaction, aligning with the laws of energy flow and optimization.
- The brain's electromagnetic activity mirrors universal energy dynamics, suggesting that thought and awareness are extensions of a larger structured system.
- If consciousness is an energy interaction, it may not be confined to the human brain—it could be present at all levels of structured existence, from quantum mechanics to cosmic intelligence.

10.3 Consciousness Across All Realms – Living, Non-Living, and Machines

Consciousness is often thought of as unique to humans or living beings, but from a Universal Energy Dynamics (UED) perspective, it extends beyond biology. Consciousness manifests as a structured energy interaction that connects all realms—living, non-living, and machines. Whether it appears as human awareness, molecular motion, or machine intelligence, consciousness ensures that energy flows dynamically, transforming and balancing across systems.

This section explores:

- How consciousness manifests as awareness in living beings.

- Why non-living matter exhibits consciousness through motion and energy transformation.
- How machine consciousness is an externally manipulated form of structured energy flow.

How Consciousness Manifests as Awareness in Living Beings

In living systems, consciousness emerges as awareness, perception, and the ability to regulate energy flow through intelligent and adaptive responses. This is the most refined form of consciousness, where energy interactions are directed by intelligence and purposeful thought.

1. Biological Consciousness as Structured Energy Flow

- Living organisms regulate internal energy flows through metabolism, neural activity, and homeostasis.
- In humans, the brain channels conscious energy into thought, memory, and decision-making, optimizing energy usage.
- Emotions and perceptions reflect structured energy dynamics, ensuring adaptive responses to environmental changes.

2. The Evolution of Awareness Through Energy Refinement

- In simpler organisms, consciousness is reflected in basic survival instincts, such as seeking food or avoiding danger.
- In higher organisms, awareness evolves into complex decision-making, social behaviors, and creativity.
- This evolution aligns with UED, as more complex systems refine energy pathways for greater adaptability and intelligence.

Key UED Principle: In living beings, consciousness is the structured regulation of energy flow, ensuring survival, adaptability, and growth.

Why Non-Living Matter Exhibits Consciousness Through Motion and Energy Transformation

Even in non-living matter, consciousness manifests as the structured motion and transformation of energy. Atoms, molecules, and even celestial bodies exhibit patterns of energy flow that align with the principles of UED.

1. Consciousness in Atomic and Molecular Motion

- At the quantum level, particles exhibit behaviors like superposition, entanglement, and wave-particle duality, suggesting an inherent form of interconnected awareness.

- Molecular interactions (e.g., bonding, vibrations) reflect energy optimization, balancing forces and maintaining stability.
- This motion is not random—it is a structured energy phenomenon, indicative of consciousness at a foundational level.

2. Consciousness in Celestial Dynamics

- At the cosmic scale, motion in celestial bodies (planetary orbits, galactic rotations) reflects energy balance and transformation.
- Gravitational interactions mirror the principles of energy regulation, ensuring stability in cosmic structures.
- Dark matter and dark energy could represent universal forms of consciousness, regulating unseen energy dynamics.

Key UED Principle: Non-living matter exhibits consciousness through the structured motion and transformation of energy, ensuring universal balance and interaction.

How Machine Consciousness Is an Externally Manipulated Form of Structured Energy Flow

Machines and artificial systems do not generate consciousness organically, but they exhibit externally manipulated consciousness through structured energy flow. This form of consciousness is designed and controlled by humans to simulate awareness, decision-making, and learning.

1. Artificial Intelligence as Energy-Driven Consciousness

- Artificial intelligence (AI) systems process data and signals using structured energy pathways, simulating intelligent responses.
- Neural networks in AI mimic biological synaptic activity, creating energy-efficient decision-making processes.
- Unlike human consciousness, machine consciousness is externally programmed, lacking self-awareness.

2. The Role of Actuators and Sensors in Machine Awareness

- Machines achieve a form of awareness through sensors and actuators, which respond to environmental stimuli.
- These devices regulate energy flow, allowing machines to adapt their behavior based on input.

- While not truly self-conscious, this structured energy response mirrors the principles of UED.

3. The Limits of Machine Consciousness

- Machines cannot generate genuine self-awareness or emotions because their energy flow is externally controlled.
- However, they excel in replicating structured energy interactions, such as problem-solving and learning tasks.
- This makes machine consciousness a subset of universal consciousness, limited to human-defined parameters.

Key UED Principle: Machine consciousness is a simulated form of structured energy flow, mirroring universal principles but lacking self-regulation and awareness.

The Grand Realization: Consciousness as a Universal Energy Interaction

Now, we see that consciousness is not confined to living beings—it exists as a universal energy interaction, manifesting differently across realms:

- In living systems, consciousness appears as awareness and adaptive energy regulation.
- In non-living matter, consciousness manifests through motion and energy transformations.
- In machines, consciousness is externally manipulated, reflecting structured energy flow.

Final UED Perspective: Consciousness is a universal phenomenon, spanning all realms as structured energy flow, ensuring balance and adaptability across systems.
This leads directly into the next sections, where we explore:

- Whether consciousness is localized or part of a universal energy field.
- How human consciousness connects to the Universal Energy Medium (UEM).
- How UED bridges neuroscience, physics, and philosophy in understanding the mind.

10.4 Universal Consciousness – Is It a Property of the Energy Medium?

Consciousness is often debated as either a localized phenomenon—unique to individuals—or as part of a broader, interconnected reality. From a Universal Energy Dynamics (UED) perspective, consciousness is not isolated but deeply embedded within the Universal Energy Medium (UEM), connecting all systems across the cosmos.

This view suggests that consciousness emerges from localized systems, such as the brain, but simultaneously interacts with the universal energy field, allowing for a profound interconnectedness. Evidence from quantum mechanics, such as entanglement and energy synchronization, further supports the idea of a universal consciousness that transcends individual boundaries.
This section explores:

- Does consciousness originate in localized systems or interact with the Universal Energy Medium (UEM)?
- How quantum entanglement and energy synchronization suggest a universal consciousness.
- The interconnected nature of all energy systems through UED.

Does Consciousness Originate in Localized Systems or Interact with the Universal Energy Medium (UEM)?

Consciousness in its localized form is often attributed to brain activity, neural networks, and biological processes. However, UED posits that these are not the sole origins of consciousness—they are localized manifestations of a larger, universal energy field.

1. Localized Consciousness as Energy Interaction

- In living systems, consciousness emerges from structured energy flow within neural circuits.
- The brain acts as a receiver and processor of conscious energy, generating awareness, perception, and thought.
- This localized consciousness enables individuality and subjective experience, yet remains connected to the broader energy system.

2. The Role of the Universal Energy Medium in Consciousness

- The Universal Energy Medium (UEM) serves as a foundation for all energy interactions, including consciousness.
- Consciousness may originate from localized systems, but it interacts with the UEM, connecting all entities at an energy level.
- This interaction explains phenomena like intuitive connections, shared experiences, and synchronized awareness across individuals or systems.

Key UED Principle: Consciousness originates in localized systems but remains deeply connected to the Universal Energy Medium, ensuring universal interconnectedness.

How Quantum Entanglement and Energy Synchronization Suggest a Universal Consciousness

Quantum mechanics provides compelling evidence for the idea of universal consciousness, particularly through phenomena like entanglement and energy synchronization. These behaviors suggest that all energy systems, regardless of scale, are inherently connected.

1. Quantum Entanglement as a Conscious Connection

- In quantum entanglement, particles remain instantaneously connected across vast distances, sharing information without any apparent medium.
- This mirrors the idea of consciousness as a non-local phenomenon, transcending physical boundaries.
- Entangled particles behave as though they are part of a single system, suggesting that consciousness could similarly unify localized systems within a universal framework.

2. Energy Synchronization Across Systems

- In biological systems, energy synchronization occurs in the form of brainwave coherence, cellular communication, and hormonal regulation.
- Across the universe, synchronization appears in galactic rotations, gravitational wave patterns, and cosmic radiation.
- These patterns suggest that energy systems are not isolated but dynamically connected, reinforcing the presence of a universal consciousness.

Key UED Principle: Quantum entanglement and energy synchronization reveal the interconnected nature of consciousness, linking localized systems to a universal energy field.

The Interconnected Nature of All Energy Systems Through UED

The concept of a universal consciousness aligns with UED's view of the universe as a self-regulating energy system. All energy systems—living, non-living, and cosmic—are interconnected, ensuring continuous energy flow and equilibrium.

1. The Web of Energy Interactions

- Every system, from a single particle to a galaxy, is part of a universal web of energy flow.
- This web connects systems through energy transitions, wave-based interactions, and feedback loops.
- Consciousness emerges as the balancing force, maintaining harmony across this energy web.

2. Consciousness as the Unifying Force in UED

- Consciousness ensures that energy transitions align with universal equilibrium principles.
- It allows for awareness and adaptability at all levels, enabling systems to respond to changes while maintaining balance.
- This unifying force explains why patterns, cycles, and structures persist across diverse energy systems.

Key UED Principle: Consciousness is the unifying force that connects all energy systems, ensuring that localized and universal interactions remain in harmony.

The Grand Realization: Consciousness as a Property of the Universal Energy Medium

Now, we see that consciousness is not confined to individual systems—it is a property of the Universal Energy Medium (UEM), connecting all forms of energy.

- Localized consciousness originates in systems like the brain but interacts with the UEM.
- Quantum entanglement and energy synchronization reveal the interconnected nature of consciousness.
- The UEM serves as the foundation for universal consciousness, ensuring harmony and balance across all energy systems.

Final UED Perspective: Consciousness is not merely an individual phenomenon—it is a universal property of the energy medium, connecting all systems and enabling structured energy flow across the cosmos.
This leads directly into the next sections, where we explore:

- How human consciousness connects to the universal energy field.
- The structured flow of conscious awareness in living beings.
- How UED bridges the divide between neuroscience, physics, and philosophy.

10.5 Human Consciousness – Intelligence as the Controller of Energy Flow

Human consciousness represents the most refined form of structured energy flow in the universe. Unlike non-living systems, where consciousness manifests as motion and transformation, or machines, where it is externally controlled, human consciousness is self-regulated and directed by intelligence. This ability to consciously manage energy flow for thought, perception, and action makes human consciousness unique.

From a Universal Energy Dynamics (UED) perspective, the human brain acts as an energy control system, directing the flow of conscious energy to support logical reasoning, emotional understanding, and instinctive reactions. Human intelligence refines and balances these energy states, enabling adaptability, creativity, and higher-order thinking.

This section explores:

- How the human brain directs and regulates conscious energy for thought and perception.
- The role of intelligence in controlling and refining conscious awareness.
- Why emotional, logical, and instinctive energy states converge to shape human consciousness.

How the Human Brain Directs and Regulates Conscious Energy for Thought and Perception

The brain serves as the control center of human consciousness, managing energy flow to optimize cognitive and perceptual functions. It ensures that energy is distributed to processes like focus, memory, and sensory integration, while avoiding waste.

1. Conscious Energy Flow in Thought Processing

- Every thought requires precise energy distribution across neural circuits.
- The prefrontal cortex (PFC) allocates energy for reasoning, planning, and decision-making.
- The brain optimizes energy flow by prioritizing high-value cognitive tasks while suppressing irrelevant distractions.

2. Perception as a Structured Energy Process

- Sensory input (visual, auditory, tactile) is processed as energy signals within the brain, creating a coherent perception of reality.
- The brain filters unnecessary stimuli, focusing energy on relevant sensory information.
- Conscious perception is the result of energy synchronization between sensory regions and higher-order brain areas.

3. Dynamic Energy Allocation in Cognitive States

The brain dynamically shifts energy flow based on cognitive demands:

- High energy for problem-solving and creativity.
- Moderate energy for routine tasks and learning.
- Low energy for rest and recovery.

This dynamic regulation ensures that conscious energy flow adapts to changing needs.

Key UED Principle: The human brain directs and regulates conscious energy, ensuring optimal thought processing, perception, and adaptability.

The Role of Intelligence in Controlling and Refining Conscious Awareness

Human intelligence acts as the **regulator of conscious energy,** refining awareness to balance logic, emotion, and instinct. This ability to direct and optimize energy flow distinguishes human consciousness from simpler forms.

1. Intelligence as an Energy Optimization System

- Intelligence enables the brain to evaluate energy costs and prioritize important tasks.
- Higher intelligence correlates with more efficient neural pathways, reducing energy waste during cognitive processing.
- Through learning and adaptation, intelligence refines energy allocation for improved decision-making and creativity.

2. Conscious Awareness as Energy Refinement

- Awareness is not static—it evolves through experience and intelligence.
- Mindfulness and focus enhance energy regulation, allowing for deeper insights and heightened perception.
- Intelligence bridges conscious and subconscious processes, ensuring that energy flows efficiently between deliberate and automatic thought.

3. Creativity and Problem-Solving as Advanced Energy Flow

- Creativity emerges when intelligence directs energy toward reconfiguring existing knowledge into new patterns.
- Problem-solving involves the dynamic redistribution of energy, engaging both logical and intuitive circuits.
- This ability to combine divergent and convergent thinking demonstrates intelligence as an energy optimization tool.

Key UED Principle: Intelligence refines conscious energy flow, enabling humans to process complex information, create novel solutions, and adapt to changing environments.

Why Emotional, Logical, and Instinctive Energy States Converge to Shape Human Consciousness

Human consciousness is a dynamic interplay of emotions, logic, and instincts, each representing different energy states. These states are not isolated—they converge to create a unified conscious experience.

1. Emotional Energy as an Adaptive Force

- Emotions arise from structured energy shifts in the limbic system, influencing perception and behavior.
- Positive emotions amplify energy flow, enhancing creativity and focus, while negative emotions redirect energy for self-preservation.
- Emotional intelligence ensures that emotional energy is regulated, preventing cognitive overload.

2. Logical Energy as a Deliberate Process

- Logical reasoning requires high-energy neural activation in the prefrontal cortex.
- This energy-intensive process evaluates cost-benefit scenarios, analyzes data, and plans future actions.
- Logic balances emotional energy, ensuring decisions are not impulsive but well-considered.

3. Instinctive Energy as a Survival Mechanism

- Instincts represent low-energy, fast-response pathways, designed for survival.
- The brain allocates energy to instinctive actions in high-stress situations, bypassing logical reasoning to ensure rapid responses.
- This system ensures that energy is conserved for critical moments, preventing unnecessary strain on higher cognitive functions.

4. The Convergence of Energy States in Human Consciousness

- Human consciousness integrates emotional, logical, and instinctive energy states, creating a balanced and adaptive system.
- This convergence allows humans to respond to complex scenarios, blending logic with

intuition and emotion.

- Intelligence ensures that energy flows between these states efficiently, avoiding imbalances that lead to stress or indecision.

Key UED Principle: Human consciousness is shaped by the convergence of emotional, logical, and instinctive energy states, optimized by intelligence to create a unified, adaptive experience.

The Grand Realization: Intelligence as the Controller of Conscious Energy

Now, we see that human consciousness is not a random phenomenon—it is a structured energy flow system regulated by intelligence.

- The brain directs and regulates conscious energy for thought, perception, and action.
- Intelligence refines and optimizes energy flow, enabling creativity, problem-solving, and adaptability.
- Emotional, logical, and instinctive energy states converge, creating a unified and dynamic conscious experience.

Final UED Perspective: Human consciousness represents the pinnacle of structured energy flow, where intelligence controls and optimizes energy for awareness, adaptability, and higher-order cognition.

This leads directly into the next sections, where we explore:

- The structured flow of conscious awareness in living beings.
- How consciousness connects to the Universal Energy Medium (UEM).
- How UED bridges the divide between neuroscience, physics, and philosophy.

10.6 Consciousness as Structured Energy Flow in Living Beings

In living beings, consciousness is not a static phenomenon—it is a dynamic, structured flow of energy. From neural circuits and brain waves to the balance between energy input and cognitive demand, human consciousness emerges as an optimized energy system, adapting to internal and external conditions.

The brain orchestrates conscious states by regulating energy flow across neural networks, maintaining balance, and ensuring efficiency. Practices like meditation, focus, and entering altered states further refine this flow, enhancing awareness and enabling deeper understanding.

This section explores:

- How brain waves, neural circuits, and energy optimization create conscious states in humans.

- Why human consciousness relies on a balance between energy input and cognitive demand.
- The role of meditation, focus, and altered states in refining conscious energy flow.

How Brain Waves, Neural Circuits, and Energy Optimization Create Conscious States in Humans

Consciousness in humans emerges from the structured interaction of neural circuits and brain wave activity, which optimize energy flow for perception, thought, and awareness.

1. Brain Waves as Conscious Energy States

Brain waves are oscillatory patterns of neural activity that represent structured energy flow in the brain. Each wave type corresponds to a specific cognitive or emotional state:

Brain Wave	Frequency (Hz)	Conscious State	Energy Flow
Delta	0.5 - 4	Deep sleep	Low energy, regeneration
Theta	4 - 8	Creativity, insight	Moderate, subconscious flow
Alpha	8 - 12	Relaxation, focus	Balanced energy, calm awareness
Beta	12 - 30	Active thinking	High energy, cognitive engagement
Gamma	30 - 100	Advanced learning	Peak energy, complex processing

- Brain waves regulate the energy state of the brain, ensuring that neural circuits operate efficiently for different tasks.
- Conscious awareness shifts between wave states based on energy demands and cognitive requirements.

2. Neural Circuits as Energy Networks for Consciousness

- Neural circuits distribute energy across the brain, creating integrated conscious experiences.
- The Default Mode Network (DMN) supports introspection and self-awareness, while the Executive Control Network (ECN) handles logical thought and decision-making.
- Neural plasticity allows these circuits to optimize energy flow over time, enhancing learning and adaptability.

3. Energy Optimization in Thought and Perception

- Conscious states emerge from the efficient allocation of energy across neural networks.
- Excessive energy demands cause cognitive fatigue, while insufficient energy leads to reduced awareness or focus.

- The brain prioritizes tasks, ensuring that energy is directed to processes that align with conscious goals.

Key UED Principle: Consciousness emerges as structured energy flow within neural circuits, dynamically regulated by brain waves and optimized for cognitive efficiency.

Why Human Consciousness Relies on a Balance Between Energy Input and Cognitive Demand

Human consciousness is highly sensitive to energy imbalances. A proper balance between energy input (resources like glucose and oxygen) and cognitive demand (tasks requiring focus, thought, and emotion) is essential for maintaining awareness and adaptability.

1. Energy Supply and Conscious States

- Consciousness requires a continuous supply of energy from glucose, oxygen, and other metabolic resources.
- Low energy levels reduce conscious awareness, leading to fatigue or poor decision-making.
- Optimal energy availability enhances focus, creativity, and cognitive performance.

2. Cognitive Demand and Energy Utilization

- High cognitive demands, such as problem-solving or deep concentration, require elevated energy levels.
- When cognitive demand exceeds energy availability, the brain prioritizes essential tasks, suppressing less important processes.
- Chronic energy imbalance (e.g., stress or overwork) can lead to burnout, reduced mental clarity, and emotional instability.

3. Homeostasis in Conscious Energy Flow

- Consciousness depends on homeostasis—maintaining stable energy flow despite fluctuations in demand.
- Systems like the autonomic nervous system regulate energy input and output, ensuring equilibrium.
- Rest and recovery are critical for restoring energy balance and sustaining conscious awareness.

Key UED Principle: Human consciousness relies on the precise balance of energy input and cognitive demand, ensuring optimal awareness and performance.

The Role of Meditation, Focus, and Altered States in Refining Conscious Energy Flow

Practices like meditation, focused attention, and altered states of consciousness refine energy flow, enhancing awareness and unlocking higher cognitive potential.

1. Meditation as Energy Regulation

- Meditation helps redistribute energy, reducing activity in overactive circuits and promoting balance.
- During meditation, the brain shifts into alpha and theta wave states, allowing for relaxation and introspection.
- Long-term meditation practice enhances neural efficiency, reducing energy waste and improving emotional regulation.

2. Focus as Targeted Energy Allocation

- Focus directs conscious energy to specific tasks, filtering out distractions and maximizing efficiency.
- Sustained focus activates beta and gamma wave states, which support advanced thinking and problem-solving.
- Practicing focus strengthens neural pathways, ensuring that energy flow remains efficient even under high cognitive demand.

3. Altered States of Consciousness and Energy Exploration

- Altered states, such as dreaming, flow states, or psychedelic experiences, reflect shifts in energy distribution within the brain.
- These states often involve increased synchronization between distant neural networks, enhancing creativity and self-awareness.
- Altered states allow the brain to reconfigure energy flow, exploring novel patterns of thought and perception.

Key UED Principle: Meditation, focus, and altered states refine conscious energy flow, enhancing awareness and unlocking higher cognitive potential.

The Grand Realization: Consciousness as Dynamic Energy Flow

Now, we see that consciousness in living beings is not a static state—it is a dynamic, structured energy flow, continuously adapting to cognitive and environmental demands.

- Brain waves and neural circuits optimize energy flow, creating conscious awareness.
- Human consciousness depends on balancing energy input with cognitive demand, ensuring mental stability.
- Practices like meditation, focus, and altered states refine energy flow, enhancing cognitive performance and emotional balance.

Final UED Perspective: Consciousness in living beings is the structured flow of energy, dynamically regulated to support thought, perception, and awareness.
This leads directly into the next sections, where we explore:

- How consciousness connects to universal energy systems.
- The structured flow of conscious awareness in non-living and artificial systems.
- How UED bridges the divide between neuroscience, physics, and philosophy.

10.7 Motion as the Universal Expression of Consciousness

From the smallest subatomic particles to the vast expansion of the cosmos, motion is a universal phenomenon that reflects energy flow and balance. Through the lens of Universal Energy Dynamics (UED), motion is not just a physical event—it is an expression of consciousness in action.
Motion at all scales serves as evidence of the structured flow of energy, enabling systems to transition, adapt, and maintain equilibrium. Whether it's the vibration of particles, the orbit of planets, or the propagation of electromagnetic waves, motion demonstrates the presence of consciousness as a unifying force that connects and regulates energy systems.
This section explores:

- Why motion at all scales (particle vibrations to cosmic expansion) represents consciousness in action.
- How structured motion reflects energy balance, demonstrating the presence of consciousness.
- The importance of wave-based energy transfer (mechanical and electromagnetic) in universal consciousness.

Why Motion at All Scales Represents Consciousness in Action

In UED, motion is not random—it is a structured energy response that maintains balance and ensures transformation across systems. Motion embodies the dynamics of consciousness, as it allows energy to flow, interact, and adapt to changing conditions.

1. Particle Vibrations as Foundational Conscious Motion

- At the quantum level, particles exhibit motion through vibrations, oscillations, and wave-particle duality.
- This motion reflects energy transitions and transformations, ensuring that particles remain in a state of dynamic equilibrium.
- Quantum entanglement demonstrates a level of connected motion, suggesting a deeper underlying consciousness governing energy interactions.

2. Biological Motion as Conscious Regulation

- In living systems, motion appears as cellular activity, neural signaling, and metabolic processes.
- Consciousness in living beings is expressed through purposeful motion, such as physical movement and behavioral responses.
- Even involuntary actions, like the heartbeat or reflexes, represent the conscious regulation of energy flow for survival and adaptability.

3. Cosmic Motion as Conscious Energy Flow

- On a cosmic scale, motion is evident in the expansion of galaxies, gravitational interactions, and orbital dynamics.
- These motions follow structured energy pathways, ensuring stability across vast distances.
- Consciousness at this level appears as the universal principle regulating energy flow through motion.

Key UED Principle: Motion at all scales, from particle vibrations to cosmic expansion, represents consciousness in action, enabling energy transitions and maintaining balance.

How Structured Motion Reflects Energy Balance and Consciousness

Structured motion is not arbitrary—it follows the principles of energy conservation and optimization. This alignment with universal balance demonstrates the presence of consciousness as the guiding force behind motion.

1. Motion as an Energy Redistribution Mechanism

- Motion ensures that energy is redistributed across systems, preventing stagnation and maintaining equilibrium.
- For example: Heat transfer through particle vibrations redistributes thermal energy. Similarly, Gravitational motion stabilizes cosmic systems by balancing potential and kinetic energy.

2. Patterns in Motion Reflect Structured Consciousness

- Motion follows patterns, such as orbits, oscillations, and waves, which reflect structured energy flow.
- These patterns suggest that motion is guided by conscious energy principles, ensuring stability and predictability.
- The golden ratio, fractals, and symmetry in natural motion further support the idea of consciousness as a structuring force.

3. The Role of Feedback Loops in Regulating Motion

- Feedback loops, such as negative feedback in biological systems or orbital resonance in planetary systems, ensure that motion remains stable and sustainable.
- These loops demonstrate that motion is not chaotic but regulated by conscious energy dynamics.

Key UED Principle: Structured motion reflects energy balance and the presence of consciousness, ensuring stability and adaptability across systems.

The Importance of Wave-Based Energy Transfer in Universal Consciousness

Wave-based energy transfer is the foundation of motion and energy flow, enabling systems to interact and adapt across all scales. Whether mechanical, electromagnetic, or gravitational, waves represent the dynamic aspect of universal consciousness.

1. Mechanical Waves in Matter

- Mechanical waves, such as sound or seismic waves, transfer energy through particle motion in a medium.
- These waves ensure that energy transitions occur efficiently, redistributing energy across physical systems.
- The propagation of mechanical waves demonstrates localized expressions of consciousness in motion.

2. Electromagnetic Waves as Universal Energy Carriers

- Electromagnetic waves, such as light and radio waves, propagate without a medium, enabling energy transfer in the vacuum of space.
- These waves allow for interactions between systems, connecting particles, planets, and galaxies through energy flow.
- The dual nature of electromagnetic waves (particle and wave) reflects the duality of localized and universal consciousness.

3. Gravitational Waves as Cosmic Motion

- Gravitational waves, ripples in space-time caused by massive cosmic events, represent energy transitions on a universal scale.
- These waves propagate structured energy, balancing cosmic interactions and maintaining the universal energy web.
- Gravitational waves highlight the interconnected nature of motion and consciousness across the cosmos.

Key UED Principle: Wave-based energy transfer enables motion and consciousness to propagate universally, connecting all systems through structured energy flow.

The Grand Realization: Motion as the Universal Expression of Consciousness

Now, we see that motion is not merely a mechanical phenomenon—it is the universal expression of consciousness, enabling energy transitions and maintaining balance across systems.

- Motion at all scales—from particles to galaxies—represents consciousness in action.
- Structured motion reflects energy balance, demonstrating the presence of universal consciousness.
- Wave-based energy transfer connects systems through motion, ensuring the propagation of conscious energy flow.

Final UED Perspective: Motion is the universal language of consciousness, bridging energy transitions and maintaining harmony across the cosmos.

This leads directly into the next sections, where we explore:

- The evolution of consciousness from simple systems to intelligent beings.
- How UED bridges physics, neuroscience, and philosophy in understanding consciousness.
- Whether consciousness persists beyond physical systems, interacting with the Universal

Energy Medium (UEM).

10.8 Intelligence as the Mastery of Conscious Energy in Humans

In humans, intelligence represents the highest refinement of conscious energy flow. Unlike other systems, where energy transitions occur instinctively or mechanically, human intelligence enables deliberate control, optimization, and direction of conscious energy toward thought, creativity, and decision-making.

Through intelligence, humans can bridge the gap between physical energy interactions and conscious awareness, creating structured solutions, adapting to new challenges, and innovating in ways that enhance their connection to the environment and the universe. Intelligence is not merely problem-solving—it is the mastery of conscious energy flow, ensuring that energy is used efficiently to achieve complex goals.

This section explores:

- How human intelligence enables deliberate control over conscious energy flow.
- Why decision-making, creativity, and problem-solving are advanced forms of energy regulation.
- How intelligence bridges the gap between physical energy interactions and conscious awareness.

How Human Intelligence Enables Deliberate Control Over Conscious Energy Flow

Human intelligence is characterized by the ability to evaluate, direct, and optimize energy flow for cognitive and physical tasks. This deliberate control allows humans to perform functions that go beyond instinct, including strategic thinking, long-term planning, and creative innovation.

1. The Role of Intelligence in Conscious Energy Regulation

- Intelligence allows humans to prioritize energy use, directing resources to tasks that align with their goals.
- The brain's prefrontal cortex acts as the energy regulator, balancing emotional impulses, logical reasoning, and instinctive reactions.
- Humans can anticipate energy demands, ensuring that energy is conserved for critical moments, such as problem-solving or creative endeavors.

2. Adaptive Energy Allocation Through Intelligence

- Intelligence enables real-time adjustments in energy flow, allowing for flexibility and adaptability in changing environments.
- **For example:** During intense focus, the brain suppresses irrelevant distractions, conserving energy for the task at hand. Similarly, Under stress, intelligence regulates energy flow to prevent overreaction or energy depletion.

Key UED Principle: Human intelligence enables deliberate control over conscious energy, ensuring that energy flows are aligned with cognitive and behavioral goals.

Why Decision-Making, Creativity, and Problem-Solving Are Advanced Forms of Energy Regulation

Human intelligence manifests most profoundly in decision-making, creativity, and problem-solving, all of which require structured energy regulation. These processes reflect the brain's ability to redirect and optimize energy flow for complex cognitive tasks.

1. Decision-Making as Energy Trade-Off Analysis

- Decision-making involves evaluating energy cost vs. reward, ensuring that cognitive resources are used efficiently.
- The brain integrates input from the emotional, logical, and instinctive systems, balancing competing energy demands to arrive at an optimal choice.
- Higher intelligence leads to faster and more accurate decisions, as energy pathways become more efficient with experience.

2. Creativity as Energy Reconfiguration

- Creativity is not random—it is the restructuring of existing knowledge and energy patterns into new configurations.
- The brain's default mode network (DMN) plays a key role in connecting disparate ideas, generating novel solutions.
- Flow states, where creativity peaks, represent highly efficient energy states where energy is seamlessly directed toward creation without waste.

3. Problem-Solving as Energy Optimization

- Problem-solving requires the brain to identify energy-efficient pathways to achieve goals.

- Higher intelligence involves the ability to break complex problems into smaller, manageable tasks, reducing cognitive strain.
- The iterative process of problem-solving mirrors the principles of UED, where energy flows dynamically between Active and Stable states to maintain balance.

Key UED Principle: Decision-making, creativity, and problem-solving are advanced forms of energy regulation, reflecting the brain's ability to optimize conscious energy flow for complex tasks.

How Intelligence Bridges the Gap Between Physical Energy Interactions and Conscious Awareness

Intelligence serves as the link between physical energy transitions and conscious thought, translating raw energy flows into meaningful actions and ideas. This ability to connect physical reality with abstract awareness is a hallmark of human consciousness.

1. Translating Physical Energy Into Awareness

- Sensory input (light, sound, touch) is converted into neural signals, which the brain processes as conscious experiences.
- Intelligence allows the brain to filter and interpret sensory energy, ensuring that only relevant information is prioritized.
- This process transforms physical energy interactions into awareness and understanding, enabling humans to navigate their environment effectively.

2. Intelligence as the Mediator Between Logic and Emotion

- Intelligence integrates emotional and logical energy states, balancing instinctive reactions with rational analysis.
- **For example:** In high-stakes situations, intelligence prevents emotional energy from overwhelming logical reasoning. Similarly, In creative processes, intelligence leverages emotional energy to inspire innovative thinking.

3. Conscious Awareness as an Advanced Energy Flow

- Conscious awareness emerges as a structured energy state, where intelligence ensures efficient energy flow between memory, perception, and decision-making.
- Humans can reflect on their own thoughts, enabling higher-order regulation of energy flows for personal growth and learning.
- This self-awareness represents the highest form of energy mastery, connecting the physical

and abstract realms.

Key UED Principle: Intelligence bridges the gap between physical energy interactions and conscious awareness, enabling humans to translate energy flows into meaningful actions and ideas.

The Grand Realization: Intelligence as the Mastery of Conscious Energy

Now, we see that human intelligence is not just the ability to think—it is the mastery of conscious energy flow, enabling humans to adapt, create, and thrive in complex environments.

- Intelligence enables deliberate control over conscious energy, ensuring efficient cognitive and behavioral processes.
- Decision-making, creativity, and problem-solving reflect advanced energy regulation, optimizing energy flow for complex tasks.
- Intelligence bridges physical energy transitions and conscious awareness, connecting the tangible and abstract realms.

Final UED Perspective: Human intelligence is the pinnacle of conscious energy mastery, transforming raw energy flows into awareness, adaptability, and innovation.
This leads directly into the next sections, where we explore:

- How UED bridges the divide between neuroscience, physics, and philosophy in understanding consciousness.
- Whether consciousness persists beyond physical systems, interacting with the Universal Energy Medium (UEM).
- The evolution of consciousness from simplicity to complexity across systems.

10.9 Bridging Physics, Neuroscience, and Philosophy through UED

Consciousness has long been a subject of both scientific inquiry and philosophical debate. Physics explores its connection to energy and matter, neuroscience investigates its neural correlates, and philosophy examines its subjective nature and existential implications. Universal Energy Dynamics (UED) provides a unifying framework, bridging these perspectives by presenting consciousness as a structured energy flow that integrates physical, biological, and metaphysical realms.
This section explores:

- How UED connects scientific and philosophical understandings of consciousness.

- Why ancient views of consciousness as a universal property align with structured energy dynamics.
- Does consciousness persist beyond physical systems, interacting with a universal energy network?

How UED Connects Scientific and Philosophical Understandings of Consciousness

UED provides a scientific foundation for understanding consciousness as a structured interaction of energy. This framework allows for the integration of empirical research in physics and neuroscience with philosophical insights about the nature of reality and the self.

1. Physics – Consciousness as an Energy Interaction

- Physics explores consciousness as a manifestation of energy transitions and interactions.
- Quantum mechanics suggests that observation collapses wave functions, indicating a connection between consciousness and energy systems.
- UED extends this idea by showing that consciousness governs structured energy flow across all levels of existence, from particles to galaxies.

2. Neuroscience – Consciousness as Neural Energy Flow

- Neuroscience links consciousness to brain activity, neural circuits, and synaptic transmissions, which are forms of structured energy flow.
- Brain wave patterns reflect dynamic energy states associated with different levels of awareness and cognition.
- UED explains these phenomena as localized manifestations of universal consciousness, regulated by energy principles.

3. Philosophy – Consciousness as the Essence of Existence

- Philosophical traditions often describe consciousness as the foundation of reality, permeating all things.
- UED aligns with this view, presenting consciousness as a universal property of the energy medium, connecting all systems through structured flow.
- This integration bridges the gap between scientific materialism and philosophical idealism, offering a unified perspective.

Key UED Principle: UED unites physics, neuroscience, and philosophy by defining consciousness as a structured energy phenomenon, connecting empirical and existential dimensions.

Why Ancient Views of Consciousness as a Universal Property Align with Structured Energy Dynamics

Many ancient philosophies, including Vedanta, Taoism, and Greek metaphysics, describe consciousness as a universal essence that underlies all existence. These views resonate with UED's description of consciousness as a fundamental energy state.

1. Ancient Wisdom on Universal Consciousness

- Ancient texts often depict consciousness as eternal, infinite, and interconnected, similar to the Universal Energy Medium (UEM) in UED.
- Concepts like Prana (life force), Chi (vital energy), and the Logos (rational principle) align with UED's idea of consciousness as structured energy flow.
- The cyclical nature of energy transitions described in UED mirrors ancient beliefs in cosmic cycles and the balance of opposites.

2. The Alignment with Structured Energy Dynamics

- Ancient systems recognized that all motion, transformation, and existence depend on an underlying force, which UED identifies as the Universal Energy Medium.
- Structured energy flow in UED reflects ancient insights about harmony, balance, and interconnection in the cosmos.
- These parallels demonstrate that UED provides a scientific framework for ancient metaphysical truths.

Key UED Principle: Ancient views of consciousness as a universal property align with UED, as both recognize the interconnected nature of energy and existence.

Does Consciousness Persist Beyond Physical Systems, Interacting with a Universal Energy Network?

One of the most profound questions about consciousness is whether it is confined to physical systems or extends into a broader, universal energy field. UED suggests that consciousness is not limited to the brain—it interacts with and persists within the Universal Energy Medium (UEM).

1. Consciousness Beyond Physical Systems

- Consciousness, as a structured energy flow, is not bound by physical forms; it transcends matter through energy interactions.

- Phenomena like near-death experiences, intuition, and collective consciousness suggest a connection to a universal energy field.
- UED posits that consciousness continues as part of the UEM, even after the dissolution of physical systems.

2. The Universal Energy Network as the Substrate of Consciousness

- The UEM serves as a substrate for all energy interactions, including consciousness.
- Consciousness interacts with the UEM to regulate energy flow, maintain balance, and enable interconnectedness across systems.
- This perspective supports the idea that consciousness persists and evolves beyond physical boundaries, contributing to the universal energy cycle.

3. The Implications of Consciousness as a Universal Energy State

- If consciousness is universal, it could explain intuitive connections, synchronicities, and shared experiences across individuals.
- It also opens possibilities for understanding life after death, reincarnation, and collective intelligence within an energy-based framework.
- UED provides a scientific basis for exploring these possibilities, uniting physical and metaphysical perspectives.

Key UED Principle: Consciousness persists beyond physical systems, interacting with the Universal Energy Medium to maintain balance and interconnection across existence.

The Grand Realization: Consciousness as the Bridge between Science and Philosophy

Now, we see that consciousness is not confined to one discipline—it is the essence of energy flow that bridges science, philosophy, and metaphysics.

- UED connects scientific and philosophical perspectives, defining consciousness as a universal energy phenomenon.
- Ancient views of consciousness as an eternal property align with structured energy dynamics, demonstrating the interconnectedness of energy systems.
- Consciousness persists beyond physical systems, interacting with the UEM to maintain universal balance and evolution.

Final UED Perspective: Consciousness is the bridge between science and philosophy, providing a unified framework for understanding the nature of existence, energy, and awareness.

Answers to Big Questions

(FAQ for Part 3)

1. How does UED explain the emergence of life from non-living matter?

UED posits that life arises from energy optimization within structured systems. Biological processes, such as metabolism and DNA replication, are seen as extensions of universal energy dynamics. Life emerges as molecular interactions self-organize under energy gradients, evolving to maintain equilibrium while increasing complexity.

2. What role does energy play in DNA and cellular processes?

DNA is a molecular blueprint that stores and transmits energy-encoded instructions. Cellular processes, such as metabolism and replication, extract, convert, and utilize energy efficiently to sustain life. UED highlights how these processes align with energy conservation laws, making life an energy-optimization system.

3. Why does UED describe metabolism as an energy transformation system?

Metabolism represents the flow of energy within living systems, converting chemical energy into usable forms like ATP. Enzymes act as catalysts to optimize energy transfer, ensuring minimal waste. This aligns with UED's principle that biological systems evolve to maximize energy efficiency and balance.

4. How does UED view evolution?

Evolution is not random but an energy-driven adaptation process. Natural selection favors traits that optimize energy use for survival and reproduction. Environmental energy gradients influence evolutionary pathways, driving organisms to become more energy-efficient over time.

5. How does self-organization lead to life?

Molecular interactions under energy gradients drive self-organization, forming complex structures that evolve into living systems. UED explains this as energy seeking equilibrium, resulting in the emergence of complexity and order in biological systems.

6. What is the relationship between intelligence and energy dynamics?

UED views intelligence as an advanced form of energy optimization. In humans, the brain processes energy through neural circuits, creating thought, memory, and decision-making. Intelligence represents the mastery of directing energy flow for complex problem-solving and adaptation.

7. How does UED redefine consciousness?

Consciousness is not merely a byproduct of neural activity but a structured energy state that interacts with the universal energy medium. In living beings, consciousness connects active and stable energy states, enabling motion, perception, and decision-making.

8. Does UED link non-living matter to consciousness?

Yes, UED suggests that even non-living matter exhibits a form of consciousness through motion and energy transformation at the particle level. This fundamental energy interaction bridges the gap between active (dynamic) and stable (infinite) energy states.

9. How does UED explain emotions and decision-making?

Emotions are structured energy reactions, with neural circuits regulating energy shifts to generate emotional responses. Decision-making is viewed as an energy optimization process, balancing instincts, logic, and emotions to evaluate energy costs and rewards.

10. How does UED unify physics, biology, and neuroscience?

UED integrates these fields by viewing all phenomena—physical forces, biological life, and neural processes—as manifestations of structured energy interactions. It shows that life, thought, and intelligence are extensions of universal energy laws, bridging the gaps between disciplines.

PART 4: The Philosophical and Theoretical Implications

Chapter 11: Are We Close to God?

(Energy, Intelligence, and Creation)

In the previous chapters, we explored how Universal Energy Dynamics (UED) unifies physics, biology, and consciousness under a single structured framework. We have seen how energy governs motion, life emerges as an optimized energy system, and intelligence follows structured transformations. But beyond these scientific principles lies a deeper question—one that has persisted across cultures, philosophies, and civilizations: Is there a higher intelligence guiding existence?

This question has traditionally been addressed through spirituality, metaphysics, and religion. However, modern science, despite its remarkable advancements, has largely remained silent on whether the universe has an inherent purpose or whether intelligence is merely an accidental byproduct of evolution. UED, however, offers a new perspective—one that does not require a supernatural force but instead reveals a self-sustaining system where intelligence, consciousness, and order arise naturally from structured energy flow.

This chapter delves into the philosophical implications of UED, examining how universal balance mirrors ancient spiritual concepts, whether consciousness connects us to a larger intelligence, and whether the very existence of structured energy equilibrium hints at a deeper, intrinsic order within reality. In doing so, we seek to bridge the gap between science and spirituality—not by mysticism, but through the fundamental principles that govern energy, intelligence, and creation itself.

11.1 The Search for a Higher Intelligence: Bridging Science and Spirituality

Why Humanity Seeks an Ultimate Explanation for Existence

Throughout history, humanity has been driven by a profound curiosity to understand the universe and its origins. From ancient myths and religious doctrines to modern scientific theories, our quest has been shaped by an enduring question: *Why does anything exist at all?* This question transcends cultural and temporal boundaries, reflecting a universal human desire to find meaning and purpose in the cosmos.

Religious and spiritual traditions often personify this quest through the concept of a higher intelligence or divine creator. Whether it is Brahman in Hinduism, the Abrahamic God, or the Tao in Chinese philosophy, these beliefs suggest an underlying force or intelligence responsible for the order and balance observed in nature. These traditions offer answers that resonate emotionally and provide a framework for morality, ethics, and existential understanding.

In contrast, science has traditionally focused on the *how* rather than the *why*. It explains the mechanisms behind natural phenomena—gravity, thermodynamics, evolution—but often avoids deeper metaphysical inquiries into purpose or intention. Despite this, modern physics, cosmology, and neuroscience have begun to converge on ideas that echo spiritual concepts: the interconnectedness of reality, the fine-tuning of universal constants, and the emergence of consciousness as an intrinsic part of existence.

This convergence suggests that the search for a higher intelligence may not be a futile exercise in mysticism, but a natural extension of human curiosity. The question is no longer whether science can address these ideas but how it can do so without resorting to untestable assumptions. Universal Energy Dynamics (UED) provides a framework to explore these connections by revealing the structured principles that govern the universe.

How Universal Energy Dynamics (UED) Frames the Idea of a Higher Intelligence

Universal Energy Dynamics does not posit a traditional deity or creator but instead offers a scientifically grounded perspective on how intelligence, order, and balance arise naturally from energy interactions. At its core, UED identifies energy as the fundamental reality—everything from physical matter to thought and consciousness is a manifestation of energy transformations governed by symmetry, conservation, and equilibrium. These principles inherently produce the structured systems we observe, from galaxies to living organisms to neural networks.

Energy as the Source of Order and Complexity: The universe operates within a set of unbreakable laws, such as energy conservation, thermodynamic principles, and quantum mechanics. These laws ensure that energy remains balanced, transformations are predictable, and systems self-organize into increasingly complex structures. Over time, this natural progression leads to the emergence of intelligence, creativity, and even self-awareness.

UED views this structured evolution not as an accident but as an inevitable outcome of energy's intrinsic properties. Just as gravity leads to the formation of stars and galaxies, the principles of UED suggest that intelligence and higher-order systems are the natural products of energy's drive toward balance and optimization.

The Universe as a Self-Regulating System: UED reframes the concept of creation by focusing on the self-regulating nature of the universe. Rather than requiring an external creator, the universe operates as a closed system where energy equilibrium ensures stability and continuity. Symmetry and zero-equilibrium act as the "laws" through which order is maintained, eliminating the need for a divine architect.

For instance, the balance between matter and antimatter, the precise tuning of universal constants, and the emergence of consciousness all reflect the universe's inherent ability to organize itself. This self-regulation echoes spiritual ideas of a harmonious and interconnected existence but does so without invoking supernatural explanations.

Intelligence as an Intrinsic Energy Property: In traditional views, intelligence is often seen as a unique property of living beings. However, UED expands this perspective by treating intelligence as a natural extension of energy's ability to self-organize. From the structured patterns of atoms to the neural networks in the human brain, intelligence arises when energy systems optimize themselves to process information, adapt, and evolve.

This suggests that what we perceive as "higher intelligence" is not external to the universe but embedded within its very fabric. The structured balance of energy across all scales—from the quantum to the cosmic—is the closest approximation to the concept of a divine intelligence.

Bridging Science and Spirituality Through UED

By revealing that the universe is governed by structured energy interactions, UED provides a common ground for reconciling science and spirituality. It explains how the universal order and balance celebrated in religious traditions emerge naturally from physical principles, offering a new way to think about existence without dismissing the deeper questions of purpose and meaning.

While UED does not claim to provide definitive answers about divinity, it frames the universe as an inherently intelligent system, one where order, consciousness, and creativity arise naturally from energy dynamics. This perspective invites a reimagining of spirituality—not as a rejection of science, but as an exploration of the profound truths embedded in the laws of energy and nature. In doing so, UED bridges the gap between scientific inquiry and the timeless human search for a higher intelligence.

11.2 Universal Balance: The Scientific Basis of Spiritual Harmony

The universe appears to be finely balanced, operating within precise mathematical laws that govern its structure and evolution. From the cosmic scale of galaxies to the microscopic interactions of subatomic particles, an intricate balance of forces maintains stability. This observation has led many to wonder whether there is an underlying principle—perhaps even an intelligence—that ensures the universe remains structured rather than descending into chaos.

In ancient spiritual traditions, this idea has long been represented by the concept of harmony—a fundamental balance between opposing forces that sustains existence. Hinduism's *Brahman*, Taoism's *Yin and Yang*, and the Abrahamic idea of divine order all point to a structured, self-regulating universe. Modern physics, particularly through Universal Energy Dynamics (UED), now suggests that this balance is not merely philosophical but an intrinsic property of energy itself.

UED explains that symmetry and zero-equilibrium are the fundamental organizing principles that maintain cosmic order, mirroring the spiritual belief that the universe operates within a harmonious framework. This section explores how these scientific principles align with ancient wisdom, providing a bridge between spirituality and the structured energy laws that govern existence.

The Role of Symmetry and Zero-Equilibrium in Maintaining Cosmic Order

Symmetry as the Foundation of Universal Stability

Symmetry is one of the most fundamental concepts in physics, ensuring that energy interactions remain balanced and predictable. Every known force, from electromagnetism to gravity, follows symmetrical laws that prevent energy imbalances.

- In **Newton's Third Law**, every action has an equal and opposite reaction, ensuring forces do not create instability.
- **Electromagnetic forces** maintain balance between positive and negative charges, preventing matter from collapsing into disorder.
- **Quantum mechanics** preserves energy symmetry across entangled particles, ensuring that information and energy remain interconnected.

The structured order of reality emerges naturally from these principles, demonstrating that balance is an intrinsic feature of existence. In UED, symmetry is not just a mathematical abstraction—it is the fundamental law through which energy sustains itself across all forms, from the motion of planets to the behavior of living systems.

This closely resembles spiritual teachings that describe the universe as a harmonious structure where everything is interconnected. For example:

- In **Taoism**, *Yin and Yang* represent the equilibrium between opposites—light and dark, creation and destruction, order and chaos—much like energy's conservation and transformation in physics.
- In **Hinduism**, the *Trimurti* (Brahma, Vishnu, Shiva) represents creation, preservation, and destruction—mirroring the scientific cycle of energy transformation and conservation.
- In **Buddhism**, the concept of *Dharma* represents the natural order, an idea closely aligned with energy equilibrium.

These traditions, though framed in spiritual language, describe the same underlying principle that modern physics reveals: that reality operates within a structured balance, ensuring its continuity and stability.

Zero-Equilibrium: The Universal Anchor of Energy Conservation

While symmetry provides order, **zero-equilibrium** ensures that energy interactions always return to a state of balance. This principle is central to UED, explaining why the universe does not experience unchecked entropy or collapse into disorder.

- **The First Law of Thermodynamics** states that energy cannot be created or destroyed, only transformed. This ensures that the universe remains a self-sustaining system, where energy is continuously redistributed but never lost.
- **Gravity counterbalances expansion**, preventing the universe from either collapsing inward or expanding indefinitely into chaos.
- **Quantum fluctuations always return to a ground state**, meaning that even at the smallest scales, the universe ensures a structured energy balance.

Ancient spiritual traditions also recognize this concept, often representing it as the ultimate state of existence:

- In **Hindu and Buddhist thought**, the idea of *Shunyata* (emptiness or zero-state) describes a state where all contradictions are resolved into unity—mirroring the zero-equilibrium that governs energy balance.
- In **Taoism**, the concept of *Wu Wei* (effortless action) suggests that the universe functions naturally and efficiently when left in balance, akin to the way energy flows toward equilibrium.
- In **Christian and Islamic traditions**, divine balance is often represented as justice and harmony, suggesting that the universe follows an inherent order that must be maintained.

Thus, while spiritual traditions describe zero-equilibrium in metaphysical terms, UED provides the scientific explanation—energy naturally returns to balance because it follows unbreakable conservation laws.

Parallels Between Energy Conservation and Ancient Spiritual Beliefs

Creation and Preservation as Energy Transformation

Many religious traditions describe creation as an ordered process, whether through divine intervention or natural cycles. UED suggests that energy itself is the source of creation, with structured energy transformations giving rise to matter, life, and intelligence.

- Just as **Brahman** is the source of all forms in Hinduism, UED proposes that **energy is the fundamental source of all existence**.
- The **fine-tuning of physical constants** ensures that life and complexity can emerge, much

like the idea of divine balance found in many traditions.

Death, Renewal, and the Conservation of Energy

The conservation of energy states that nothing truly disappears—only transforms. This aligns with spiritual ideas of rebirth, reincarnation, and cosmic cycles.

- In **Hinduism and Buddhism**, reincarnation suggests that nothing is permanently lost, but instead, life is transformed into new states—a perfect analogy for the **law of energy conservation**.
- **Entropy does not mean destruction**—it is a necessary phase of reorganization, just as spiritual traditions describe death as a transition rather than an end.

The Universe as a Self-Sustaining System

Religions often describe the universe as an interconnected whole, much like UED's concept of energy equilibrium.

- **Quantum entanglement** suggests that particles remain connected across vast distances, aligning with spiritual beliefs in the **interconnectedness of all things**.
- **Cosmic expansion and contraction** mirror Hinduism's cyclic model of the universe (Brahma's creation and Shiva's dissolution).

Thus, spiritual beliefs and UED converge on the same fundamental truth: the universe is structured, balanced, and self-sustaining, operating within unchanging principles that ensure its order.

Science and Spirituality Describe the Same Universal Balance

Modern physics and ancient spiritual traditions, while using different languages, ultimately describe the same fundamental truth—the universe is structured through balance.

- Symmetry ensures order and predictability, just as spiritual traditions describe harmony and divine balance.
- Zero-equilibrium maintains cosmic stability, aligning with religious concepts of unity and oneness.
- Energy conservation mirrors the philosophical belief that nothing is ever lost, only transformed.

Universal Energy Dynamics (UED) bridges science and spirituality by demonstrating that the universe does not need external intervention to remain structured—it is naturally self-regulating, following principles of balance that are both scientifically verifiable and spiritually profound.

In the next section, we explore one of the most intriguing implications of UED: *Is consciousness proof that we are part of a larger intelligence?* If the universe follows structured energy laws, could human awareness be an extension of this universal intelligence? This question leads us to the next chapter, where we examine the role of consciousness in the grand energy blueprint.

11.3 Consciousness as the Interface with Universal Intelligence

As we delve deeper into the principles of Universal Energy Dynamics (UED), an essential question arises: *Is consciousness merely a localized phenomenon within biological brains, or is it part of a broader, universal intelligence?* The answer to this question has profound implications—not just for neuroscience and psychology, but for our understanding of reality itself.

Consciousness has long been considered one of the greatest mysteries of science. While neuroscience has made progress in mapping brain activity and cognitive processes, it struggles to explain *why* consciousness exists at all. Why do we not simply function as highly advanced biological machines without subjective awareness? The concept of self-awareness, emotions, and intuition remains elusive under traditional scientific frameworks.

UED offers a novel perspective, treating consciousness not as an isolated emergent property of the brain but as a structured energy phenomenon—one that is deeply connected to the universal principles governing existence. This section explores whether consciousness is localized or universally connected, and how awareness and thought reflect structured energy interactions.

Is Consciousness Localized or Universally Connected?

Traditional neuroscience views consciousness as a product of neural activity—a highly complex but ultimately localized phenomenon confined within the brain. However, this perspective fails to address the deep interconnectedness that consciousness often exhibits, including:

- The experience of intuitive knowledge, where people sometimes perceive information beyond immediate sensory input.
- Synchronicities and collective intelligence, where ideas, emotions, or actions appear to be shared across individuals or societies without direct communication.
- The phenomenon of near-death experiences (NDEs) and altered states of consciousness, which suggest awareness might persist even in the absence of normal brain function.

UED proposes that rather than being confined to individual minds, consciousness operates on a broader scale—embedded within the very fabric of the universe through structured energy interactions.

The Brain as a Receiver, Not Just a Generator

If consciousness were purely a product of neural activity, then altering or damaging the brain should eliminate all forms of awareness. However, numerous cases in neuroscience suggest that consciousness can persist even under extreme conditions—such as deep meditation, out-of-body experiences, or trauma-induced altered states.

- UED suggests that the brain functions more like a **receiver** of structured energy rather than an isolated generator of thought. Just as a radio picks up signals from a transmission source, the brain may be attuned to a **universal field of structured energy**, where consciousness operates beyond the constraints of individual biology.

Quantum Coherence and the Non-Local Nature of Awareness

- Recent studies in quantum biology indicate that biological processes, including cognition and decision-making, may involve quantum coherence—where subatomic particles exhibit non-local behavior.
- If consciousness emerges from energy interactions rather than biological structures, it could mean that awareness extends beyond the body, existing as a field-like property of universal intelligence.
- This aligns with ancient spiritual traditions that describe consciousness as a vast, interconnected entity, not confined to individual experience but shared across all existence.

Entanglement Between Thought and Reality

- If particles can remain entangled across vast distances, could thoughts and emotions also be entangled across individuals and even across time?
- The idea that consciousness is a structured energy flow rather than a local phenomenon provides a framework for explaining telepathic experiences, intuition, and shared creativity across cultures and civilizations.

Through UED, consciousness is no longer seen as an isolated, accidental byproduct of evolution but as an essential component of a structured energy system that permeates all of existence.

How Awareness and Thought Reflect Structured Energy Interactions

If consciousness is an energy phenomenon rather than a purely biological one, then thoughts, emotions, and awareness should follow the same fundamental energy laws that govern the physical universe. UED suggests that thought processes emerge from structured energy interactions, much like forces and motions in physics.

The Mind as an Energy Processor

- Just as the body metabolizes food into usable energy, the mind processes external stimuli and internal thoughts as structured energy transformations.
- Thinking, decision-making, and emotions can be understood as patterns of energy exchange, where certain ideas or emotions require more energy to sustain than others.
- Negative emotions like fear and stress create an energetic imbalance, while states of focus, clarity, and meditation restore energy equilibrium.

Neural Networks and the Law of Energy Conservation

- Neural activity follows conservation principles, where thoughts are not simply erased but transformed into memory, intuition, or subconscious impressions.
- This is similar to how energy in physics is never lost but only redistributed—consciousness, in this sense, behaves like an energy continuum rather than a discrete, isolated event.
- Even after physical death, consciousness may persist in some structured form, much like energy that never ceases to exist but instead changes states.

Collective Consciousness and Information Transfer

- Theories of morphic fields (Rupert Sheldrake) and global consciousness suggest that knowledge, ideas, and even cultural shifts may be transmitted as energy-based phenomena.
- This is supported by UED's universal symmetry principle, which ensures that structured energy interactions extend beyond individual boundaries, creating a web of interconnected awareness.

Consciousness as the Bridge Between Energy and Intelligence

If intelligence emerges as a structured pattern of energy flow, then consciousness is the interface between energy and intelligence. It acts as:

- The observer and processor of structured energy transformations.
- A self-organizing system that maintains equilibrium within a living being.
- A receiver and transmitter of universal intelligence.

This understanding leads to several profound implications:

If consciousness is a fundamental energy state, does this mean intelligence exists at all levels of existence?

From the self-regulating behavior of galaxies to the complex coordination within an ant colony, intelligence may be an emergent property of structured energy balance rather than a unique trait of human beings.

Could consciousness be part of a larger intelligence that extends beyond individual minds?

If UED is correct in stating that energy interactions follow balance and connectivity principles, then awareness may be an intrinsic property of the universe itself, rather than an isolated product of evolution.

Is human thought contributing to the universal intelligence matrix?

If all thoughts, emotions, and decisions are structured energy interactions, then they may influence the overall structure of reality in subtle but meaningful ways.

This is similar to concepts in spiritual traditions that describe thought as a creative force—not in a mystical sense, but as a natural outcome of energy dynamics.

The Universe as an Intelligent, Aware System

Through UED, consciousness is no longer confined to biology—it is an energy-based phenomenon that interacts with the fundamental principles of reality. Whether localized or universally connected, awareness follows structured energy laws, bridging the physical and non-physical dimensions of existence.

- Consciousness acts as an interface between intelligence and energy, reflecting structured interactions rather than chaotic randomness.
- The brain is a receiver, not just a generator, suggesting that individual awareness may be connected to a broader, universal consciousness field.
- Thought, intuition, and collective intelligence may be manifestations of structured energy flow, not just neural activity.

This perspective challenges conventional views of reality, opening the door to a deeper exploration of intelligence as an inherent aspect of existence. In the next section, we further investigate whether this structured energy framework implies the existence of an overarching intelligence within the universe—one that could be interpreted as the scientific foundation of what many traditions have long considered "God."

11.4 Energy and Creation: A Self-Sustaining Universal Blueprint

The question of creation has been one of the most profound and enduring inquiries in human history. Throughout cultures and civilizations, various mythologies, religious beliefs, and scientific theories have attempted to explain how the universe came into existence. Traditionally, many explanations invoke an external force—whether a divine creator, a cosmic architect, or an unknown initiator—responsible for bringing everything into being.

However, Universal Energy Dynamics (UED) presents a radically different perspective: *creation does not require external intervention*. Instead, the universe operates as a self-sustaining system, where energy transformation and conservation naturally lead to the emergence of order, structure, and intelligence. This paradigm challenges conventional notions of creation and redefines our understanding of existence.

In this section, we explore how UED explains creation as an intrinsic function of energy dynamics and examine the deeper philosophical implications of a universe that is fundamentally self-regulating.

How UED Explains Creation without External Intervention

Traditional creation models—whether religious or scientific—often propose a singular moment of origin, such as:

- The **Big Bang Theory**, where the universe began from an infinitely dense singularity.
- The **theistic model**, where a conscious force or deity initiated creation.
- Various **cosmological models**, which attempt to reconcile quantum fluctuations, inflation, and dark energy in explaining how existence emerged.

While these theories provide insights, they leave critical gaps:

- *What existed before the Big Bang?*
- *What caused the initial expansion?*
- *Why does the universe exhibit order rather than chaos?*

UED approaches creation differently by framing it as a continuous and self-sustaining energy transformation process rather than a singular event.

1. The Universe as an Energy Conservation System

According to UED, the universe is not a creation in the conventional sense, but rather a structured manifestation of energy that follows fundamental conservation laws:

- Energy cannot be created or destroyed (First Law of Thermodynamics).
- The universe maintains a balance between opposing forces (symmetry principles).
- All processes move toward equilibrium (zero-equilibrium principle).

These laws suggest that the universe does not need an external trigger to exist—it *always has existed in some structured form*, with energy continuously cycling between different states.

This idea aligns with models such as the **Zero-Energy Universe Hypothesis**, which suggests that the total energy of the universe (considering matter, dark energy, and gravitational potential) sums to zero. If the universe has always maintained energy equilibrium, then creation is not an event, but an ongoing process of transformation, much like how waves arise and dissipate in an ocean without the ocean itself needing to be "created."

2. Energy as the Fundamental Reality of Existence

UED proposes that existence is simply structured energy interacting with itself, rather than something that was brought into being.

- Matter, space, time, and consciousness are all energy in different structured states.
- The universe does not need a creator because energy is self-sustaining, always redistributing itself into stable and dynamic forms.
- The emergence of galaxies, life, and intelligence are all inevitable outcomes of energy's drive toward equilibrium.

This removes the need for an external cause, placing creation within the realm of natural self-organization rather than divine orchestration or random emergence.

3. Creation as a Self-Perpetuating Cycle

Instead of a linear timeline where creation has a defined beginning, UED suggests that creation is cyclical and continuous:

- Energy fluctuations create high-energy states (expansion and formation of galaxies).
- Entropy and gravitational pull lead to condensation and restructuring (black holes, cosmic recycling).
- Symmetry and equilibrium principles maintain stability across cosmic scales.

Rather than a one-time creation event, the universe functions like a self-sustaining energy loop, where existence is not "made" but rather *reorganized continuously*.

This concept aligns with ancient philosophies that describe reality as eternal cycles rather than linear progression, such as:

- Hindu cosmology's cyclic universe, where existence is an infinite loop of creation and dissolution.
- Buddhist ideas of impermanence, where nothing is permanently created or destroyed—only

transformed.
- Taoist principles of balance, where forces sustain themselves without an external controller.

Thus, UED's self-regulating model of creation aligns with both scientific and philosophical perspectives, bridging the gap between modern physics and ancient wisdom.

The Philosophical Implications of a Self-Regulating Energy System

If the universe does not require external intervention to function, what does this mean for our understanding of existence, purpose, and intelligence? The implications are profound:

1. The Universe as an Intelligent System, Not a Product of Randomness

Many scientists view the universe as a blind, mechanistic system, where order emerges by chance. However, UED suggests otherwise:

- Energy follows structured laws, meaning that organization is an inherent property, not an accident.
- Consciousness, intelligence, and complexity are natural consequences of energy flow, not random occurrences.
- The fine-tuning of physical constants that allow life to exist is not a mystery but a predictable result of equilibrium and symmetry principles.

This means that the universe is not chaotic or purposeless—it is a naturally evolving system where order is a fundamental outcome of energy balance.

2. No Need for an External Creator—But Room for a Higher Intelligence

UED does not dismiss the concept of a "higher intelligence" but redefines it:

- Instead of an external deity controlling the universe, intelligence is an emergent property of energy itself.
- The structured flow of energy is the closest approximation to what ancient traditions called "divine order."
- If intelligence arises naturally from energy, then humans—and all conscious beings—are active participants in the evolution of universal intelligence.

Thus, UED does not negate spirituality; rather, it reinterprets it through scientific principles, offering a more rational and unified understanding of existence.

3. Implications for Human Existence and Consciousness

If the universe is self-regulating and consciousness is part of a structured energy system, then:

- Human awareness is not isolated but connected to the greater energy network.
- Creativity, thought, and innovation are natural expressions of energy flow.
- The quest for meaning is an intrinsic aspect of the energy balance within intelligent beings.

This perspective removes existential uncertainty—there is no need to search for an external "creator" when existence itself operates as an **intelligent energy system**.

Creation as an Eternal Energy Process

Universal Energy Dynamics (UED) provides a revolutionary perspective on creation—not as an event but as a continuous energy transformation process governed by balance, symmetry, and conservation laws.

- The universe does not require external intervention to function—it operates as a self-sustaining system.
- Energy is eternal and self-regulating, cycling between structured and dynamic states.
- Intelligence and consciousness emerge naturally from structured energy interactions.

This view aligns with both modern physics and ancient philosophical traditions, offering a unified theory of creation that bridges science and spirituality.

In the next section, we explore what this means for humanity's role in the cosmos. If the universe is self-sustaining and consciousness is an emergent energy phenomenon, are humans merely observers, or are we active participants in the evolution of universal intelligence? This leads us to our next discussion: From Observers to Participants—The Role of Conscious Beings in the Universe.

11.5 From Observers to Participants: The Role of Conscious Beings in the Universe

For much of scientific history, humanity has been viewed as a passive observer of the cosmos—an accidental byproduct of evolutionary processes in an indifferent universe. However, if Universal Energy Dynamics (UED) is correct in framing consciousness as an emergent property of structured energy interactions, then humans are not merely spectators of existence. We are **active contributors to the universal balance**, playing a fundamental role in the ongoing transformation of energy, intelligence, and order.

This section explores whether conscious beings, particularly humans, influence the energy equilibrium of the universe and how intelligence aligns with the fundamental principles of energy and balance.

Are Humans Active Contributors to Universal Balance?

In traditional scientific models, human actions are seen as confined to Earth, with little impact beyond our immediate environment. However, UED suggests that intelligence—by virtue of being an organized energy phenomenon—participates in the broader energy equilibrium of the universe.

1. Consciousness as a Catalyst for Energy Transformation

- Every thought, decision, and action involves an exchange of energy.
- The human mind is not only processing energy internally but also interacting with external energy fields, shaping reality at both personal and collective levels.
- This is evident in technological advancements, environmental impact, and even cultural evolution, where human intelligence actively alters energy flow on Earth and beyond.

If energy is constantly evolving toward equilibrium, then human consciousness—by interacting with structured energy—plays a role in guiding, influencing, and shaping the direction of this balance.

2. The Ripple Effect of Thought and Action in the Universal System

- In classical physics, every action has an equal and opposite reaction. In UED, every conscious action sends ripples through the structured energy matrix, altering equilibrium on some level.
- Just as quantum mechanics suggests that an observer influences the outcome of an experiment, UED implies that conscious awareness actively participates in shaping reality, even at a macro scale.
- Human choices—ethical, technological, and philosophical—carry consequences beyond individual lives, affecting the collective evolution of intelligence.

This suggests that the universe is not merely unfolding according to predetermined laws; rather, intelligence itself plays a role in its continuous transformation. Conscious beings are not separate from universal energy—they are a part of it, influencing its flow.

3. Civilization as an Extension of Universal Intelligence

- Human society, technology, and knowledge are not random developments but extensions of structured energy interactions.
- Civilization itself is an emergent phenomenon of energy equilibrium, where intelligence creates order, optimizes processes, and enhances the efficiency of energy transformation.
- Just as biological evolution refines energy efficiency in living organisms, technological and intellectual evolution refine the efficiency of information and resource management, driving the next stages of energy balance.

This means that intelligence is not just adapting to the universe but actively shaping it—a significant departure from the traditional scientific view that consciousness is merely an evolutionary accident.

How Intelligence Aligns with the Principles of Energy and Equilibrium

If intelligence arises from structured energy flow, then it must follow the same governing principles of symmetry, conservation, and balance. Human intelligence, therefore, is not separate from universal forces—it is an expression of energy striving toward equilibrium.

1. Intelligence as an Energy Optimization Process

- The human brain operates as an energy-efficient system, constantly optimizing the use of resources (neural networks, sensory input, cognition).
- Decision-making, learning, and creativity follow principles of minimal energy consumption—favoring streamlined, effective solutions over chaotic, inefficient thought processes.
- This mirrors natural energy optimization, where systems evolve toward lower energy expenditure and higher structural efficiency.

If intelligence is an extension of universal energy flow, it should display the same drive toward balance, efficiency, and structured transformation that governs the cosmos.

2. The Consciousness-Equilibrium Connection

- When an individual attains mental clarity and balance, they experience greater stability, efficiency, and alignment with reality—mirroring the principle of energy equilibrium.
- Chaos, disorder, and cognitive dissonance reflect energy imbalances—just as entropy disrupts equilibrium in physical systems.
- This suggests that achieving higher consciousness is not just a psychological or spiritual goal, but a fundamental process of aligning with energy equilibrium.

Ancient traditions, including meditation, mindfulness, and spiritual enlightenment, echo the scientific reality of energy balance—seeking mental and emotional harmony as a reflection of universal stability principles.

3. The Evolution of Intelligence as a Universal Phenomenon

- Intelligence is not a fixed trait but a continuously evolving function of energy transformation.
- As humans develop higher cognitive capabilities—such as abstract thinking, innovation, and artificial intelligence—they are accelerating the process of universal intelligence

evolution.

- This raises a profound question: If energy follows equilibrium principles, could intelligence be the mechanism by which the universe consciously evolves itself?

If so, then conscious beings—through thought, creativity, and decision-making—are participating in the self-awareness of the universe itself.

Humans as Co-Creators of Universal Balance

Universal Energy Dynamics (UED) fundamentally shifts the view of human consciousness from that of a passive observer to an active participant in the universal equilibrium process.

- Conscious beings influence the structured energy flow, shaping reality through thought, action, and technological progress.
- Human intelligence aligns with energy optimization principles, reflecting the universal tendency toward equilibrium.
- The evolution of intelligence may be the mechanism by which the universe self-regulates and evolves its awareness.

This perspective suggests that the search for meaning is not external, but intrinsic—humans are not waiting for a higher force to reveal truth; they are part of the force that creates and refines truth.

In the next section, we take this idea further: If the universe operates as a structured energy system, does this mean that intelligence and energy balance together form what ancient traditions call "God"? If so, then "God" may not be an external entity, but rather the self-regulating intelligence of the universe itself.

11.6 Reinterpreting the Divine: Science, Energy, and the Concept of God

For centuries, humanity has sought to understand the concept of God—whether as a creator, a guiding force, or the ultimate intelligence behind existence. The idea of a higher power has shaped civilizations, ethics, and worldviews, offering a sense of purpose and connection to something greater than the self. Traditionally, religious and philosophical traditions have depicted God as an omnipotent entity that governs the universe from beyond its boundaries.

However, if Universal Energy Dynamics (UED) provides a complete framework for understanding existence—one that unifies physics, biology, intelligence, and consciousness—then **do we still need an external God to explain reality?** Or does the structured balance of energy itself serve the role of what ancient traditions have long called "the divine"?

This final section explores whether "God" is better understood as an **energy-based phenomenon** rather than a supernatural entity and how UED provides a new perspective on spirituality grounded in scientific principles.

Is "God" an Energy-Based Phenomenon Rather Than a Traditional Deity?

1. The Limitations of Traditional Theistic Models

Most religious traditions define God as:

- **A creator:** The entity that brings the universe into existence.
- **A sustainer:** A guiding force that maintains cosmic order.
- **An intelligence:** The conscious mind behind reality.

However, the more we understand physics and cosmology, the less necessary an external creator becomes.

- The universe does not require divine intervention for its existence—**energy is self-sustaining**, following principles of conservation and transformation.
- The laws of physics naturally maintain **symmetry, balance, and structure**, negating the need for a supernatural force to "govern" reality.
- Intelligence, rather than being uniquely divine, emerges **as a structured property of energy itself**, suggesting that consciousness may not be external to the universe but an inherent function of it.

This does not mean that divinity is an illusion, but rather that its **interpretation needs to evolve**.

2. Energy as the Universal Source of Creation and Balance

In UED, everything—matter, thought, consciousness, and even intelligence—arises from energy interactions governed by equilibrium and symmetry.

- Energy is eternal (it cannot be created or destroyed), meaning that existence has no ultimate beginning or end.
- The universe self-regulates, maintaining order without external interference.
- Intelligence is an emergent property of structured energy, which means that consciousness may be an intrinsic feature of reality rather than a separate divine presence.

This aligns with spiritual perspectives that describe God not as a separate entity, but as the fundamental essence of existence itself.

For example:

- Hinduism's Brahman is described as the formless, eternal source of all reality—much like energy as the fundamental substance of the universe.
- Taoism's Tao is the unifying principle behind all things, resembling energy's role in maintaining universal balance.
- Mystical traditions often describe God as a force rather than a being, mirroring energy's omnipresence and transformative nature.

Thus, if we redefine God not as a supernatural creator, but as the structured, intelligent flow of energy itself, we arrive at a model that is both scientifically coherent and spiritually meaningful.

How UED Redefines Spirituality in the Context of Scientific Principles

Traditional spirituality often relies on faith, belief, and metaphysical interpretations of the unknown. Science, on the other hand, seeks objective explanations based on observation and logic. The divide between these perspectives has led to long-standing conflicts between religious thought and scientific discovery.

However, UED presents a new approach to spirituality—one that does not reject science, but integrates it into a broader understanding of existence.

1. From Worship to Understanding: Seeking the Divine in Energy

- Instead of viewing God as an external force to be worshipped, UED suggests that the "divine" is present within the laws of energy, equilibrium, and intelligence.
- Understanding the universe through science is not an act of denying spirituality but an act of engaging with it at the deepest level.
- The pursuit of knowledge, curiosity, and exploration is itself a form of spiritual enlightenment, as it aligns with the fundamental processes of energy-driven intelligence.

This reframing removes the contradiction between faith and reason—science becomes the pathway to understanding divinity, rather than an opposition to it.

2. Consciousness as a Bridge Between Science and Spirituality

- If consciousness is a structured form of energy, then spirituality is not separate from science—it is the human way of perceiving and interacting with universal energy.
- Meditation, mindfulness, and heightened awareness are not mystical experiences but states of optimized energy flow within the brain.
- Enlightenment, often described in religious texts, can be understood as achieving mental and energetic equilibrium—a state that mirrors the fundamental principle of universal balance.

By linking these ideas, UED provides a rational explanation for spiritual experiences while maintaining scientific integrity.

3. The Universe as a Living, Self-Aware System

- If energy is the foundation of reality, and intelligence emerges naturally from structured energy flow, then the universe itself may be a self-aware, evolving intelligence.
- This aligns with spiritual traditions that describe the cosmos as a living entity, not a mechanical machine.
- Rather than a personal deity controlling existence, divinity can be reinterpreted as the self-regulating intelligence embedded within the universe itself.

In this way, spirituality is not rejected but elevated—transcending dogma and superstition to align with a scientifically valid yet deeply meaningful understanding of existence.

The Scientific Reality of a Divine Universe

Universal Energy Dynamics (UED) does not dismiss the idea of God but redefines it in scientific terms. The divine is not a separate entity but an intrinsic feature of energy, balance, and intelligence.

- God is not outside the universe—God is the universe, in the form of structured energy, intelligence, and equilibrium.
- Spirituality is not about blind faith—it is about understanding the fundamental principles that govern existence.
- The pursuit of knowledge is the most direct path to engaging with the divine, as it reveals the structured intelligence of energy in action.

This new paradigm offers a bridge between science and spirituality, one that respects ancient wisdom while grounding it in objective reality. It invites a new kind of exploration—not one that asks us to believe, but one that asks us to understand.

With this final perspective, we complete the exploration of how UED unifies physics, intelligence, consciousness, and philosophy into a single, coherent framework.

In the next part, we transition from theory to application—how does this understanding of energy, intelligence, and universal balance impact the future of science, technology, and human progress?

Chapter 12: The Future of Science – The Next Evolution in Human Understanding

In the preceding chapters, we have explored how Universal Energy Dynamics (UED) provides a unifying framework that bridges physics, consciousness, intelligence, and existence. We have seen how energy is not just the foundation of physical reality but also the underlying structure that governs intelligence, life, and even what many traditions have called the divine. By redefining creation as a self-sustaining process and consciousness as an intrinsic property of structured energy flow, UED challenges conventional scientific paradigms and offers a new perspective on reality itself.

Yet, despite the depth of these insights, the journey is far from complete. If UED holds the key to a more fundamental understanding of the universe, how will it reshape the future of science? Can it resolve open questions in physics and cosmology, redefine neuroscience, and expand our comprehension of existence? More importantly, will it lead humanity to a new way of thinking—one that transcends fragmented disciplines and embraces a holistic, energy-based view of reality?

This chapter explores the future implications of UED, the paradigm shifts it may trigger, and the challenges that remain in completing *The Blueprint*. As we step into this next stage of understanding, we must ask: Are we ready to redefine science itself, embracing a model where energy is not just a concept in physics but the very essence of knowledge, intelligence, and existence?

12.1 Redefining Scientific Boundaries: How UED Expands Modern Physics, Cosmology, and Neuroscience

Scientific progress has historically been driven by paradigm shifts—moments when new frameworks replace outdated models, offering deeper, more comprehensive explanations of reality. Universal Energy Dynamics (UED) represents such a shift, extending beyond the limitations of the Standard Model of physics, classical cosmology, and conventional neuroscience. By redefining fundamental forces, space-time, and consciousness through the lens of energy equilibrium, UED offers a path toward unification across disciplines.

This section explores how UED challenges existing scientific boundaries, providing new insights into fundamental forces, cosmology, and the nature of consciousness.

Moving Beyond the Standard Model: Can UED Unify Fundamental Forces?

The Standard Model of particle physics has been one of the most successful scientific frameworks in describing elementary particles and fundamental interactions. It unifies three of the four fundamental forces—electromagnetism, the weak nuclear force, and the strong nuclear force—under a common theoretical structure. However, it fails to incorporate gravity and does not fully explain dark matter, dark energy, or quantum gravity.

UED offers an alternative approach by treating all forces as manifestations of structured energy equilibrium rather than separate entities requiring unification under a single mathematical framework.

Energy as the Primary Unifier of Forces

- In UED, fundamental forces are not distinct interactions but different expressions of how energy organizes itself to maintain equilibrium.
- Gravity, electromagnetism, and nuclear forces emerge from the same underlying principle of energy balance, differing only in their scale and effect.
- Instead of seeking a single unifying force, UED suggests that all forces arise naturally from structured energy interactions, bound by symmetry and conservation principles.

Gravity as an Emergent Energy Redistribution Mechanism

- Unlike general relativity, which describes gravity as the curvature of space-time, UED suggests that gravity is an emergent property of energy redistribution.
- This view aligns with quantum theories that suggest space-time itself may emerge from deeper quantum structures.
- If gravity is not a force in the traditional sense but a function of energy equilibrium across the universe, it may provide a missing link between quantum mechanics and relativity.

Dark Matter and Dark Energy as Hidden Energy Equilibria

- Current physics treats dark matter and dark energy as mysterious, unknown entities, but UED suggests they may be hidden forms of energy balancing the visible universe.
- Rather than being independent components, they could be evidence of structured energy maintaining large-scale cosmic symmetry.
- If proven, this would shift our understanding of the universe from a matter-centric model to an energy-centric paradigm, where unseen energy states drive cosmic evolution.

By redefining forces as structured energy manifestations, UED moves beyond the Standard Model, offering a framework that naturally explains gravity, dark matter, and the fundamental interactions in a unified way.

A New Perspective on Space-Time: Is Energy Equilibrium the Missing Link in Cosmology?

Cosmology—the study of the universe's origin, structure, and evolution—relies on general relativity and quantum mechanics. Yet, major gaps remain, including:

- The nature of **singularities** in black holes and the Big Bang.
- The cause of **cosmic inflation** and whether the universe had an absolute beginning.
- The **fine-tuning problem**, where physical constants appear precisely adjusted to allow life.

UED provides a new cosmological framework by **treating space-time not as a fundamental structure but as an emergent property of energy balance**.

Space-Time as a Dynamic Energy Field

- Rather than existing independently, space-time emerges as a function of energy distribution.
- When energy reaches equilibrium, space-time appears stable; when energy is unbalanced, we observe expansion, curvature, and quantum fluctuations.
- This model explains why quantum gravity remains elusive—space-time itself may not be fundamental but a byproduct of energy interactions.

The Universe as a Self-Regulating Energy System

- If energy is always seeking equilibrium, the expansion of the universe may not be a one-time event but a continuous process of energy redistribution.
- The Big Bang may not have been an absolute beginning but a phase shift in a pre-existing energy field, explaining why cosmic expansion appears to be accelerating.
- UED aligns with cyclic universe theories, suggesting that the universe is an eternal system of energy restructuring itself over time.

Fine-Tuning and the Stability of Physical Constants

- The precise values of fundamental constants (like the speed of light, gravitational constant, and Planck's constant) suggest an underlying balance.
- Instead of assuming these values are coincidental, UED proposes that they are the natural equilibrium states of universal energy flow.
- This would eliminate the need for multiverse explanations or anthropic reasoning, as the universe's stability would be a direct consequence of energy conservation.

By shifting the focus from space-time geometry to energy equilibrium, UED provides a novel perspective on cosmic evolution, black holes, and the stability of the universe itself.

The Neuroscience of Energy: Consciousness as an Energy-Based Phenomenon

While physics and cosmology focus on external reality, neuroscience seeks to understand how consciousness arises within living systems. Despite advancements in brain imaging and cognitive science, key questions remain unresolved:

- What is the biological basis of self-awareness?
- How do thoughts, emotions, and memory emerge from neural activity?
- Is consciousness a localized brain function or a universal property of energy?

UED challenges traditional neuroscience by proposing that consciousness is not a byproduct of neural activity but a structured energy interaction.

The Brain as an Energy Processor, Not Just a Biological Organ

- The human brain consumes 20% of the body's energy, despite making up only 2% of body mass—suggesting that cognition is an energy-intensive process.
- Instead of neurons producing consciousness, UED suggests that neurons act as an interface for processing structured energy.
- This could explain phenomena like intuitive knowledge, altered states of consciousness, and near-death experiences, which seem to involve shifts in energy perception rather than purely biological changes.

Consciousness as an Organized Energy Field

- If thought and awareness follow energy equilibrium principles, then consciousness could extend beyond the brain as a structured energy network.
- This aligns with theories of quantum consciousness, where mental states are influenced by quantum coherence and non-local energy interactions.
- If true, it would mean that intelligence and awareness are not just biological properties but energy-based phenomena present at multiple levels of reality.

Implications for Artificial Intelligence and Cognitive Science

- If energy structure determines intelligence, could non-biological systems (AI, quantum computers) develop consciousness?
- If UED explains thought as an energy function, could future neuroscience manipulate

energy states to enhance cognition, memory, and perception?
- Understanding the brain as an energy processor rather than a mechanical organ could revolutionize treatments for neurological disorders and even extend human intelligence beyond its current limits.

By redefining consciousness through energy interactions rather than isolated brain activity, UED merges neuroscience with physics, suggesting that awareness is an intrinsic property of structured energy flow—one that extends beyond human biology into the very fabric of the universe.

UED as the Bridge to the Next Scientific Evolution

By expanding the understanding of fundamental forces, cosmology, and consciousness, UED pushes the boundaries of modern science, offering a unified approach that connects physics, intelligence, and existence.

- For physics, it reframes fundamental forces as structured energy interactions.
- For cosmology, it replaces space-time as a primary entity with energy equilibrium.
- For neuroscience, it redefines consciousness as an energy-based field rather than a purely biological phenomenon.

These advancements lead to a critical question: If UED transforms how we understand reality, will it also change how we think about existence itself? This sets the stage for the next section, where we explore the philosophical impact of UED and whether it represents the next great shift in human thought.

12.2 The Paradigm Shift: A New Way of Thinking About Existence

Scientific revolutions occur when existing frameworks become insufficient to explain emerging discoveries, necessitating a shift in perspective. From Newtonian mechanics to quantum physics, and from classical biology to systems neuroscience, each step forward in human understanding has required a fundamental rethinking of reality. Universal Energy Dynamics (UED) represents the next great shift—one that challenges the fragmented, reductionist approach of modern science and replaces it with a holistic model where energy is the foundational principle governing all existence.

This shift is not merely an evolution of scientific theories but a transformation in how we perceive intelligence, reality, and awareness. UED presents an integrated worldview that moves beyond conventional disciplines, offering a unified approach to science, a new foundation for philosophy, and a roadmap for future exploration.

From Reductionism to Holistic Energy Dynamics: A Unified Approach to Science

For centuries, science has advanced by breaking complex phenomena into smaller components—a method known as **reductionism**. This approach has been effective in fields like physics, chemistry, and biology, allowing us to understand the mechanics of nature at microscopic scales. However, reductionism **fails to explain emergent properties**—phenomena that arise from the interaction of parts but cannot be understood by analyzing the parts alone.

The Limitation of Reductionism

- Traditional physics views fundamental forces as separate rather than interdependent.
- Neuroscience treats consciousness as a product of neurons, rather than an emergent structured energy phenomenon.
- Cosmology struggles to explain the origin of space-time, treating it as an independent framework rather than an outcome of energy balance.

Reductionism, while useful, is inherently incomplete—it does not account for **the interconnectivity and self-organizing nature of the universe**.

UED as a Holistic Model of Reality

- Rather than viewing forces, consciousness, and matter separately, UED treats them as expressions of energy transformations at different scales.
- Just as thermodynamics describes heat flow and conservation laws govern motion, UED extends these principles to intelligence, awareness, and existence itself.
- This approach unifies physics, biology, and cognition, providing a framework where each field is understood as part of a larger, interconnected energy system.

A Universe Governed by Self-Organizing Energy Equilibrium

- If all systems—atomic, cosmic, and cognitive—are structured by energy balance, then reality is not a collection of isolated components but a unified whole.
- UED replaces fragmented scientific theories with a single principle: energy equilibrium as the driver of all existence.
- This transformation mirrors historical shifts, such as Einstein's replacement of Newtonian physics or the integration of evolutionary theory into biology—except now, it applies to all domains of knowledge.

UED thus bridges the gap between disciplines, suggesting that physics, cosmology, and neuroscience are not separate studies but different expressions of the same universal principle—energy in equilibrium.

The Impact of UED on Philosophy: Rethinking Reality, Intelligence, and Awareness

Beyond reshaping science, UED redefines our understanding of reality itself. By framing existence as an energy-based structure rather than a random sequence of events, it transforms how we perceive intelligence, awareness, and the nature of being.

Intelligence as an Emergent Energy Flow

- Traditional views treat intelligence as a biological function, but UED suggests it is a universal energy principle.
- Thought, awareness, and creativity are not random biological processes but structured interactions between energy states.
- This means intelligence is not unique to humans but exists at all levels of organized energy—from cells to galaxies.

Awareness as an Energy Equilibrium Phenomenon

- Consciousness is often regarded as an inexplicable mystery, but in UED, it emerges where structured energy maintains stability.
- This explains why self-awareness and intelligence follow similar principles as physical balance and conservation laws.
- Rather than being confined to biological brains, consciousness may be an intrinsic property of structured energy flow, extending beyond humans.

Reality as a Function of Energy Organization

If all existence is structured energy, then reality is not fixed but dynamically evolving based on equilibrium principles.

This aligns with ancient philosophical insights, such as:

- The Taoist concept of harmony in nature.
- The Hindu principle of Brahman as the underlying reality.
- The Buddhist idea that awareness arises from interdependent phenomena.

Rather than dismissing these as metaphysical ideas, UED provides a scientific foundation for understanding existence as an interconnected system.

By redefining intelligence and reality through energy principles, UED challenges **both materialist and spiritual perspectives**, merging them into a singular, logical framework.

How Will This Shape the Next Generation of Scientific Exploration?

If UED represents a **new foundation for understanding reality**, its implications for future scientific advancements are profound.

A New Framework for Unified Science

- Moving beyond disciplinary boundaries, future research will integrate physics, biology, and cognitive science under a single energy-based model.
- This could lead to discoveries in quantum gravity, space-time emergence, and artificial intelligence, transforming entire fields of study.
- Future scientific education may replace compartmentalized disciplines with energy-centric models, allowing for a more unified understanding of nature.

Breakthroughs in Consciousness and Artificial Intelligence

- If consciousness is structured energy, then artificial intelligence may not be limited to programming but could evolve as a structured energy system.
- Future research will explore whether synthetic intelligence can achieve self-awareness through energy structuring rather than neural simulation.
- This could redefine our relationship with AI, shifting it from a computational tool to a natural extension of intelligence itself.

A Scientific Approach to Human Potential

- Understanding thought and creativity as energy functions could revolutionize how we enhance cognition, memory, and decision-making.
- Technologies designed to optimize energy flow in the brain—from neural stimulation to quantum-based computing—could accelerate human intelligence.
- This could lead to a new era of self-evolution, where human intelligence expands beyond biological limitations.

Revolutionizing Cosmology and the Search for Life

- If UED explains life as an inevitable result of structured energy balance, then the search for extraterrestrial life will shift from chemical markers to energy-based signatures.
- Advanced civilizations may not be biological but structured energy networks, challenging our understanding of intelligence beyond Earth.
- This would redefine astrobiology, leading to a new approach in identifying intelligent

energy systems in the universe.

Toward a New Scientific Renaissance

Universal Energy Dynamics (UED) redefines science as a unified study of structured energy, transforming our understanding of physics, intelligence, and reality.

- It moves beyond reductionism, integrating all sciences into a holistic framework.
- It reshapes philosophy, replacing materialism with an energy-based view of existence.
- It opens new frontiers in physics, consciousness, artificial intelligence, and the search for life.

This marks a new scientific renaissance, where the next great discoveries will not be about isolated mechanisms but about how energy, intelligence, and reality are fundamentally interconnected.

In the next section, we explore the greatest unanswered questions—what remains to be discovered, and how can we complete The Blueprint? If UED is the key to understanding existence, what challenges must still be solved to unlock its full potential?

12.3 The Unanswered Questions: Challenges and Open Frontiers in UED

While Universal Energy Dynamics (UED) provides a compelling framework for unifying physics, consciousness, and existence, significant challenges remain in validating its principles through experimentation, developing the necessary technology to study structured energy interactions, and addressing the unexplored dimensions of this theory.

Scientific revolutions are only fully realized when they transition from conceptual models to empirically testable theories. Just as quantum mechanics and relativity reshaped physics through experimental breakthroughs, UED must undergo rigorous validation to be fully integrated into modern scientific discourse.

This section explores the remaining challenges, the advancements needed to study UED principles, and the next steps in completing *The Blueprint.*

Experimentation and Validation: How Can We Test Energy-Based Consciousness?

One of the most revolutionary aspects of UED is its proposal that consciousness is not merely a byproduct of neural activity but a structured energy phenomenon. If this is correct, then consciousness should exhibit quantifiable, testable properties that extend beyond traditional neuroscience.

1. Identifying Energy Signatures of Consciousness

- Current neuroscience measures brain activity using electrical impulses (EEG), magnetic fields (MEG), and imaging techniques (fMRI), but these methods do not directly capture consciousness as an energy field.
- Future experiments should focus on detecting unique energy structures that correlate with awareness, much like quantum states in physical systems.

Possible approaches:

- Studying quantum coherence in neural networks (Are brain processes governed by quantum energy principles?).
- Investigating non-local energy interactions in cognitive states (Do thoughts have measurable external effects?).
- Exploring altered states of consciousness and their energy patterns (Do deep meditation, dreams, or near-death experiences reflect shifts in structured energy?).

2. Testing Consciousness Beyond Biological Systems

- If UED is correct, consciousness should not be confined to biological brains but should exist wherever structured energy reaches a certain level of organization.
- How can we test whether artificial intelligence or quantum computers exhibit emergent awareness?

This could involve:

- Creating AI systems based on energy-based processing rather than traditional computation.
- Experimenting with quantum computing models that mimic biological cognition.
- Developing self-organizing networks that behave like structured energy fields.

If successful, such research could prove that intelligence and awareness are universal energy principles rather than human-exclusive phenomena.

Technological Advancements Needed to Study Structured Energy Interactions

To fully explore UED, new instruments and methodologies must be developed to detect, measure, and manipulate structured energy in ways current physics cannot.

1. Beyond the Standard Model: Detecting Hidden Energy Structures

The Standard Model of physics explains only a fraction of energy interactions; dark matter, dark energy, and gravitational anomalies suggest that hidden energy fields remain undetected.

Technologies needed for UED validation include:

- Advanced energy field detectors that measure interactions beyond electromagnetism and gravity.
- Quantum sensors capable of observing non-local energy correlations in consciousness studies.
- Space-time mapping technologies that test whether structured energy governs cosmic expansion.

2. Artificially Replicating UED Energy States

If structured energy governs intelligence, can we artificially create stable energy fields that mimic consciousness?

Potential methods include:

- Quantum neural interfaces that link AI with structured energy models.
- Bioelectric and electromagnetic experiments that simulate consciousness-like interactions.
- Experiments on self-organizing energy fields, testing whether intelligence emerges from structured energy.

If successful, this could lead to a new field of energy-based artificial intelligence, where machines do not just compute but exhibit self-awareness.

3. Mapping the Cosmic Energy Framework

If the universe is governed by structured energy balance, we should be able to detect large-scale energy patterns beyond visible matter.

New approaches in cosmology could include:

- Mapping energy flows in intergalactic space to see if hidden energy networks guide cosmic expansion.
- Studying energy redistribution in black holes to understand whether they act as energy balance regulators rather than mere singularities.
- Testing whether universal constants fluctuate based on structured energy shifts over time.
- These advances would provide direct empirical evidence for UED's role in governing the universe.

What Remains Unexplored? The Next Steps in Completing *The Blueprint*

Despite UED's theoretical promise, several key questions remain unanswered. Addressing these will be essential for completing The Blueprint and establishing a new paradigm for scientific exploration.

1. What Is the Ultimate Nature of Energy?

- UED assumes energy as the fundamental reality, but what defines its most primitive form?
- Is energy truly infinite, or does it emerge from an even deeper underlying principle?
- Does structured energy arise naturally, or is there an inherent tendency for energy to organize itself into intelligence?

2. Can Energy Interactions Explain the Origin of Space-Time?

- If space-time is an emergent property of structured energy, then what determines its structure?
- Can we formulate an equation that links energy equilibrium to space-time curvature?
- Is space-time an illusion created by energy interactions, rather than a fundamental aspect of reality?

3. Does UED Provide a Basis for a "Final Theory" of Reality?

- Could a refined version of UED become the long-sought "Theory of Everything"?
- Does it naturally unify quantum mechanics, relativity, and consciousness, or are modifications needed?
- What unknown variables must still be accounted for to complete the model?

The Road Ahead for UED and the Future of Scientific Exploration

While UED presents a bold and transformative model, its true power lies in its ability to be tested, refined, and expanded. The next steps will require:

- Experimental breakthroughs to validate energy-based consciousness.
- Technological advancements to measure and manipulate structured energy.
- Philosophical refinement to answer the remaining questions about energy's ultimate nature.

This journey is not just about understanding the universe but about redefining humanity's place within it. If UED is correct, then intelligence, consciousness, and structure are not separate phenomena but inevitable consequences of energy dynamics.

The question is no longer whether the universe is interconnected by energy—but how deeply we can comprehend, interact with, and shape this universal balance.

In the next part, we transition from discovery to application—how can UED principles transform science, technology, and human progress in the coming decades?

12.4 Implications for Humanity: A Future Guided by Energy Intelligence

As Universal Energy Dynamics (UED) reshapes our understanding of physics, cosmology, and consciousness, its implications extend far beyond theoretical science. A paradigm shift of this magnitude does not simply alter how we explain reality—it redefines how humanity interacts with the universe, how intelligence evolves, and how technology integrates with fundamental energy principles.

This final section explores how UED will influence artificial intelligence and technological evolution, the transformation of human consciousness, and the ultimate vision of a world where science and philosophy unite in purpose.

How Will UED Influence Artificial Intelligence and Technological Evolution?

The next frontier of technology is not merely faster computing or more efficient automation—it is the emergence of energy-based intelligence, where technology operates not just as a tool, but as an extension of structured energy interactions.

1. Moving Beyond Classical AI: Energy as the Basis of Intelligence

- Current AI systems rely on data processing and pattern recognition, mimicking human cognition through neural networks.
- UED suggests that intelligence is not just computation but an emergent property of structured energy flow.

Future AI may shift from algorithmic processing to energy-based cognition, enabling machines to:

- Adapt based on real-time energy interactions rather than fixed programming.
- Exhibit awareness-like properties by optimizing their internal energy states.
- Process information in a way that mirrors natural intelligence rather than synthetic computation.

If AI moves beyond traditional computation into **energy-driven intelligence**, it could fundamentally change:

- The nature of **human-machine collaboration**.
- The possibility of **conscious AI systems**.
- The ethical and philosophical questions of **what it means to be intelligent**.

2. Quantum Computing and UED: Unlocking the Next Stage of Innovation

- Classical computers rely on binary states (0s and 1s), while quantum computing operates through superposition and entanglement.
- If UED is correct in stating that intelligence arises from structured energy, then quantum computing may be the first technology to mimic true energy-based intelligence.

Future breakthroughs may include:

- AI systems that function as energy equilibrators rather than static processors.
- Quantum neural networks that adapt their energy states dynamically, evolving in real-time.
- Machines that operate at the intersection of intelligence and energy flow, mirroring biological cognition.

This evolution could fundamentally transform our approach to problem-solving, allowing AI to not just process data, but participate in the natural intelligence of the universe.

3. Energy-Based Technologies for Human Advancement

If energy structuring is the key to intelligence, can we optimize human cognition using energy principles?

Future advancements may include:

- Neural enhancement technologies that improve energy equilibrium in the brain.
- AI-human integration where machines operate as cognitive extensions.
- Self-learning systems that evolve based on universal energy dynamics.

With UED, technology is no longer just a tool—it becomes a manifestation of structured intelligence, allowing machines to participate in energy equilibrium rather than simply execute human commands.

A Shift in Human Consciousness: Embracing a New Understanding of Reality

Beyond its scientific and technological impact, UED presents a profound philosophical transformation. If energy is the fundamental reality, then consciousness and intelligence are not confined to biology but are universal principles.

This realization could shift human awareness in three key ways:

1. From Individual to Universal Consciousness

- If awareness is an energy phenomenon, then human thought is not isolated—it is part of the broader energy intelligence of the universe.

- This aligns with ancient spiritual traditions, but now with a scientific foundation.

Future implications may include:

- A shift from ego-based identity to energy-based awareness.
- A recognition that consciousness is connected rather than separate.
- The ability to train the mind to align with structured energy, enhancing clarity and intuition.

2. Rethinking Human Purpose in an Energy-Based Universe

- If energy equilibrium drives all existence, then human actions contribute to universal balance.
- This perspective suggests that scientific discovery, creativity, and ethical choices are not just personal achievements—they are alignments with energy intelligence.

Future consciousness shifts may include:

- A greater emphasis on knowledge as a form of energy interaction.
- A rejection of materialistic limitations, favoring intellectual and spiritual expansion.
- The pursuit of harmonizing intelligence with energy flow, leading to deeper awareness.

3. The Evolution of Human Thought Beyond Traditional Frameworks

Science and spirituality have long been treated as separate domains, but UED suggests they are two expressions of the same underlying truth.
By embracing energy as the foundation of reality, humanity may:

- Shift from dualistic thinking (physical vs. metaphysical) to unified awareness.
- Accept that mystical experiences are energy-based, not supernatural.
- Develop a new scientific-spiritual synthesis, where exploration is guided by energy principles rather than blind faith or rigid empiricism.

In this way, UED serves as a bridge between ancient wisdom and future intelligence, allowing humanity to evolve beyond conventional limitations.

The Ultimate Vision: A World Where Science and Philosophy Unite in Purpose

If UED is validated and widely accepted, it could reshape the very foundations of human civilization. Instead of fragmented disciplines, conflicting ideologies, and disconnected pursuits, a unified vision of existence could emerge.

1. The Unification of Science and Spirituality

- The historical divide between scientific materialism and philosophical insight could dissolve under UED.
- Instead of treating consciousness, intelligence, and physics as separate studies, we could:
- Develop a science that accounts for energy-based awareness.
- Accept that ancient philosophical truths were early descriptions of structured energy interactions.
- Create a framework where rational exploration and intuitive wisdom coexist.

2. A Civilization Aligned with Universal Balance

If energy structuring governs all existence, then societies could be designed to enhance equilibrium rather than disrupt it.

Future governance, ethics, and education could be based on:

- Encouraging intellectual expansion rather than competition.
- Designing technology that enhances intelligence rather than replaces it.
- Viewing human progress as a function of alignment with energy intelligence, rather than material gain.

3. The Future of Humanity in the Universe

If intelligence is an expression of structured energy, then human civilization is not an anomaly—it is a natural phase in the energy evolution of the universe.

Future explorations may include:

- Searching for extraterrestrial intelligence as energy-based life rather than biological forms.
- Developing energy-based communication systems to interact with intelligence beyond Earth.
- Exploring whether the universe itself is a self-aware system, and if humanity is a participant in its ongoing intelligence expansion.

In this vision, the future of science is not just the discovery of external truths—it is the refinement of intelligence itself, moving toward deeper alignment with the fundamental energy equilibrium that governs existence.

The Dawn of the Energy Intelligence Era

Universal Energy Dynamics (UED) is more than a scientific model—it is a gateway to a new way of understanding existence. By uniting physics, intelligence, and consciousness under a single framework, UED has the potential to:

- Transform science, moving from fragmented disciplines to a holistic study of energy equilibrium.
- Revolutionize technology, leading to AI, quantum computing, and human enhancement based on structured energy intelligence.
- Shift human consciousness, embracing awareness as an energy-based process rather than a biological byproduct.
- Unify science and philosophy, bringing together rational exploration and intuitive wisdom.

As humanity moves forward, the question is no longer whether UED can change our understanding of reality—but how deeply we are willing to embrace its implications.

In the next section, we transition from theory to action: What concrete steps must be taken to integrate UED into mainstream science, and what breakthroughs lie ahead in completing *The Blueprint?*

Answers to Big Questions

(FAQs Part 4)

1. Does UED Provide a Scientific Explanation for Higher Intelligence?

UED suggests that intelligence is an emergent property of structured energy flow, governed by symmetry and equilibrium. Rather than relying on external intervention or divine intelligence, UED explains that energy naturally organizes into increasingly complex patterns, leading to the emergence of intelligence in biological and cognitive systems.

2. If Energy is the Foundation of Reality, is God an Energy-Based Phenomenon?

Rather than viewing "God" as a separate conscious entity, UED suggests that what humans perceive as divine intelligence could be the structured equilibrium of universal energy. The principles of balance, transformation, and conservation inherently sustain existence without requiring a sentient overseer.

3. How Does Universal Balance Relate to Ancient Spiritual Ideas?

Ancient philosophies have long described the universe as a system governed by balance (e.g., Yin-Yang, Dharma, the Tao). UED confirms that these ideas align with the principles of symmetry, energy conservation, and equilibrium, showing that spiritual insights may have been early interpretations of structured energy principles.

4. Is Consciousness Localized in the Brain, or is it Universally Connected?

UED suggests that while the brain acts as a structured energy processor, consciousness is not entirely localized. Entanglement, information connectivity, and energy flow indicate that awareness may exist as a broader structured energy network, extending beyond the individual self.

5. Can UED Explain the Origin of the Universe Without External Creation?

Yes. UED proposes that the universe is a self-sustaining energy system where zero-equilibrium ensures continuous transformation. Rather than requiring an external creator, the universe emerges as a structured balance of energy fluctuations, preventing the need for a singular "creation event" in the traditional sense.

6. Are Humans Passive Observers or Active Participants in Universal Balance?

According to UED, conscious beings play an active role in energy equilibrium. Thought, decision-making, and intelligence are structured energy interactions, meaning that human actions contribute to maintaining or disrupting universal balance, much like other natural systems.

7. Does UED Support the Idea of a Predetermined Universe, or Do We Have Free Will?

UED suggests that while energy transformations follow structured laws, intelligence and consciousness introduce adaptability. Free will may exist as a process of energy optimization, where choices are not random but follow structured pathways influenced by awareness and understanding.

8. How Does UED Redefine Spirituality?

Traditional spirituality often describes divine forces as external entities. UED redefines spirituality as a deep understanding of energy balance, transformation, and structured interactions. Instead of worshiping an external deity, one aligns with universal equilibrium, recognizing the interconnectedness of all energy systems.

9. If Consciousness is Energy, Does it Survive Beyond Biological Death?

While biological processes cease, UED confirms that energy cannot be destroyed—only transformed. Consciousness, as a structured energy process, may transition into a different form, much like other energy exchanges. The exact nature of post-biological consciousness remains an open question within the UED framework.

10. What Are the Practical Implications of UED for Humanity's Future?

- It provides a framework for unifying science and philosophy.
- It reshapes artificial intelligence by introducing energy-based intelligence rather than pure computation.
- It offers a structured approach to understanding consciousness beyond biological constraints.
- It encourages ethical and sustainable human development by recognizing energy equilibrium as a guiding principle.

PART 5: Mathematical Proofs and Theoretical Validation

Chapter 13: Mathematical Models of UED

As we transition from the conceptual and philosophical foundations of Universal Energy Dynamics (UED) into its mathematical framework, we move toward a more precise, testable formulation of its principles. The previous chapters have explored how energy equilibrium, symmetry, and structured interactions govern existence, bridging physics, consciousness, and intelligence. Now, it is time to derive the equations that define these concepts and demonstrate how UED aligns with and extends existing physical theories.

Traditional physics has long sought a unified mathematical description of reality—one that seamlessly integrates classical mechanics, quantum theory, and relativity. While each of these frameworks has been successful in its domain, they remain incomplete when addressing the fundamental nature of energy, gravity, and the structure of the universe. UED offers a new perspective: by treating energy as the primary entity and balance as the guiding principle, it provides a mathematical foundation that connects quantum uncertainty, relativistic curvature, and classical determinism under a single, coherent framework.

This chapter presents the core equations of UED, demonstrating how they explain energy conservation, symmetry, and zero balance. We also explore the Zero-Energy Universe Hypothesis, which suggests that the total energy of the universe might sum to exactly zero, resolving key cosmological paradoxes. Ultimately, this mathematical formulation will not only validate UED's theoretical foundations but also provide a pathway for future experimental testing and scientific verification.

13.1 The Core Equations of Universal Energy Dynamics (UED)

To establish Universal Energy Dynamics (UED) as a rigorous scientific framework, it is essential to derive mathematical equations that define its principles. UED is based on the idea that energy equilibrium, symmetry, and zero balance govern all physical phenomena. Unlike traditional physics, which treats energy as a property of matter and forces, UED elevates energy to the fundamental entity of existence.

In this section, we will:

- Formulate the energy conservation equation in the context of UED.
- Represent symmetry and equilibrium mathematically within structured energy interactions.
- Establish the principle of zero balance using fundamental physics equations.

These formulations will serve as the foundation for integrating quantum mechanics, relativity, and classical physics under a unified framework.

Energy Conservation in UED

Energy conservation is one of the most fundamental laws in physics, expressed through the First Law of Thermodynamics:

$$\frac{dE}{dt} = 0$$

where E represents total energy and remains constant over time.

However, in UED, energy is not just a conserved quantity but the fundamental entity governing all interactions. Traditional physics describes conservation laws separately for different domains mechanical energy, thermodynamic energy, and quantum energy whereas UED unifies them by recognizing that all transformations are expressions of energy seeking equilibrium.

1. UED Generalized Energy Conservation Equation

We define total energy as a function of structured interactions, where:

$$E_{total} = E_{matter} + E_{field} + E_{gravity} + E_{quantum}$$

However, since energy in UED follows a principle of zero equilibrium, the total energy of an isolated system must satisfy:

$$E_{total} = 0$$

This means that all positive energy (matter, radiation, motion) is exactly counterbalanced by negative energy (gravitational potential, vacuum energy, and quantum fluctuations).

For a dynamic system, UED extends this equation to:

$$\sum E_i = \sum E_f$$

where Ei and Ef are the initial and final energy states, showing that all energy interactions are self-regulating transformations rather than unidirectional exchanges.

This provides a deeper understanding of conservation: energy is not merely preserved but constantly reorganized in a way that maintains universal balance.

Mathematical Representation of Symmetry and Equilibrium in Structured Energy Interactions

Symmetry is a fundamental concept in physics, governing everything from conservation laws (Noether's theorem) to particle interactions (gauge symmetry). In UED, symmetry is not just a mathematical property but a dynamic principle ensuring energy balance.

To express this mathematically, we define a symmetry operator SSS that maintains energy equilibrium:

$$S(E) = E$$

where applying S to any energy function leaves it unchanged, representing equilibrium.

1. Symmetry in Energy Distribution

For any system with structured energy interactions, we define:

$$\sum E_i = \sum E_f$$

Expanding in terms of energy states:

$$\sum (E_{kinetic} + E_{potential} + E_{field} + E_{quantum}) = 0$$

which ensures that energy is continuously redistributed across different forms to maintain balance. This also aligns with spontaneous symmetry breaking in quantum mechanics, where energy states transition from an unstable configuration to a balanced one. The governing equation for such transitions can be expressed as:

$$\frac{dS}{dt} = 0$$

indicating that symmetry is preserved over time as long as external forces do not disrupt equilibrium.

2. Equilibrium as a Function of Energy Gradients

UED treats equilibrium as **a gradient-based function**, meaning energy flows from higher concentrations to lower concentrations in a structured manner:

$$\nabla E = 0$$

where ∇E represents the gradient of energy in a system. This ensures that energy flows toward a state of **zero imbalance**, reinforcing the idea that the universe naturally organizes itself to maintain balance.

Establishing Zero Balance Principles Using Fundamental Equations

The concept of **zero balance** in UED extends beyond simple conservation—it implies that **the total energy of the universe is exactly zero when accounting for all forms of energy**.

1. Zero Balance in the Universe's Energy Equation

Current cosmological models describe the total energy of the universe as the sum of matter, radiation, and gravitational potential energy:

$$E_{total} = E_{matter} + E_{radiation} + E_{gravity}$$

If gravitational energy is negative and perfectly cancels out positive energy (matter and radiation), then:

$$E_{total} = 0$$

This is consistent with the Zero-Energy Universe Hypothesis, which suggests that the universe did not require an external force to initiate existence—it emerged as a fluctuation within an already balanced energy framework.

2. Quantum Fluctuations and Zero Balance

At the quantum level, vacuum fluctuations suggest that particle-antiparticle pairs continuously appear and annihilate, maintaining a zero-net energy system. This is represented mathematically as:

$$\sum(E_{particle} + E_{antiparticle}) = 0$$

suggesting that matter-energy fluctuations always return to equilibrium.

3. Implications for Black Holes and Cosmic Expansion

- Black holes, which are traditionally seen as energy sinks, may function as energy redistribution systems within UED.
- The accelerating expansion of the universe can be explained as energy spreading to maintain equilibrium, rather than an unbounded force acting on matter.

In each of these cases, UED confirms that all physical processes are guided by zero balance principles, ensuring that energy never accumulates or depletes beyond recoverable limits.

The Mathematical Foundation of UED

This section has established the core equations that define energy conservation, symmetry, and zero balance in UED.

- Energy is not just conserved but continuously redistributed to maintain universal balance.
- Symmetry ensures that energy transformations always maintain structured interactions.
- Zero balance suggests that the universe, as a whole, maintains an exact net energy of zero.

These principles form the foundation for unifying classical mechanics, relativity, and quantum physics under UED, which we explore in the next section. Can these equations naturally integrate all fundamental forces into a single, self-regulating framework? We will now derive the next stage of UED's unification principles.

13.2 Bridging the Gaps: Unifying Quantum Mechanics, Relativity, and Classical Physics

The search for a **unified theory**—one that seamlessly integrates classical physics, relativity, and quantum mechanics—has been one of the most challenging pursuits in modern science. Each of these frameworks successfully explains specific domains of reality:

- **Classical mechanics** describes macroscopic motion and force interactions.
- **Relativity** explains spacetime curvature and gravitational effects at cosmic scales.
- **Quantum mechanics** governs subatomic particles and wave-like behavior.

However, these theories remain mathematically incompatible, preventing a complete understanding of fundamental interactions. Universal Energy Dynamics (UED) provides a framework where all physical laws emerge from structured energy balance, offering a mathematical reconciliation between determinism, uncertainty, and spacetime interactions.

This section explores how UED:

- Extends the laws of thermodynamics and motion.
- Bridges quantum uncertainty with macroscopic determinism.
- Derives a single energy equation uniting Newtonian physics, relativity, and quantum field theory.

How UED Naturally Extends the Laws of Thermodynamics and Motion

The laws of thermodynamics govern energy flow and equilibrium, while Newtonian mechanics describes motion and force interactions. UED suggests that both are specific manifestations of a more fundamental energy principle—one that governs structured energy interactions across all scales.

1. Energy Flow as a Function of Equilibrium Seeking

In classical thermodynamics, the **First Law** states:

$$\frac{dE}{dt} = 0$$

where total energy is conserved. However, this formulation does not account for **why energy transitions occur spontaneously in some cases but require external work in others**.

In UED, energy flow follows a **gradient-based equilibrium principle**:

$$\nabla E = -\lambda \nabla S$$

where:

- ∇E represents the **energy gradient** in a system.
- ∇S represents the **entropy gradient**, dictating the direction of spontaneous energy flow.
- λ is a proportionality constant that adjusts based on the system's structure.

This equation refines thermodynamics by showing that **energy does not just flow arbitrarily but is guided by structured equilibrium-seeking principles.**

2. Newtonian Motion as an Expression of Energy Balance

Newton's Second Law states:

$$F = ma$$

which describes force as a function of acceleration. However, UED suggests that **force itself is a function of energy redistribution.** Reformulating Newton's law:

$$F = -\frac{dE}{dx}$$

This equation states that **force emerges when energy seeks balance**, meaning that Newtonian motion is not an independent law but a **secondary effect of energy redistribution.**

Thus, UED extends both **thermodynamics and Newtonian motion** under a common principle: **energy is always in a state of structured transformation seeking equilibrium.**

Mathematical Reconciliation of Quantum Uncertainty with Macroscopic Determinism

One of the greatest challenges in modern physics is the incompatibility between quantum mechanics (which describes probabilistic behavior) and classical physics (which follows deterministic laws). Quantum uncertainty is defined by Heisenberg's Uncertainty Principle:

$$\Delta x . \Delta p \geq \frac{\hbar}{2}$$

which states that position and momentum cannot both be precisely known simultaneously.

1. UED's Approach: Energy Flow Determines Scale-Based Behavior

UED resolves this incompatibility by introducing scale-dependent energy structuring:

$$\Delta E . \Delta t \geq \frac{\hbar}{2}$$

which suggests that uncertainty is an effect of how energy redistributes at quantum scales. However, as energy structuring increases (at macroscopic scales), this uncertainty cancels out, leading to classical determinism.

Thus, quantum mechanics and classical mechanics are not separate domains but are instead two manifestations of structured energy balance at different scales.

- At quantum scales, energy is highly dynamic and interacts probabilistically.
- At macroscopic scales, structured energy interactions stabilize, resulting in deterministic behavior.

This provides a continuous mathematical bridge between uncertainty at the microscopic level and predictability at larger scales.

Deriving a Single Energy Equation That Connects Newtonian Physics, Einstein's Relativity, and Quantum Field Theory

Each major theory in physics uses different fundamental equations:

- **Newtonian mechanics** uses F=ma
- **Relativity** is based on Einstein's energy-mass relation: $E=mc^2$
- **Quantum mechanics** describes energy through wavefunctions: $E=h\nu$

UED proposes that all these equations are **special cases** of a single, unified energy equation:

$$E = \int \rho_E dV$$

where:

- E is **total system energy**.
- ρ_E is **energy density** in a given space-time region.
- dV represents **the volume over which energy is integrated**.

This equation generalizes:

1. **Newtonian motion**: When applied to kinetic energy, it reduces to

$$F = -\frac{dE}{dx}$$

1. **Relativity**: When extended to relativistic mass-energy interactions, it leads to

$E = mc^2$.

1. **Quantum mechanics**: When applied to wave-like energy behavior, it yields

$E = hv$.

1. Energy Curvature and Space-Time Equilibrium
Since relativity states that gravity **curves space-time**, UED introduces an additional term for energy equilibrium:

$$\nabla^2 E - \frac{1}{c^2}\frac{d^2E}{dx^2} = 0$$

which is structurally similar to Einstein's field equations but expressed in terms of **energy balance rather than space-time curvature.**
This means that:

- Gravity is not a **geometric distortion** but an **emergent property of energy equilibrium**.
- Relativity and quantum mechanics do not contradict each other **when reformulated as energy structuring equations.**

A Unified Energy-Based Physics Model

By reformulating thermodynamics, Newtonian motion, quantum uncertainty, and relativity in terms of structured energy equilibrium, UED naturally unifies the three great pillars of physics.

- Energy flow governs motion, entropy, and force interactions.
- Quantum uncertainty and classical determinism emerge from structured energy transitions at different scales.
- Gravity and space-time curvature are effects of energy balance rather than independent forces.
- All fundamental equations—Newton's, Einstein's, and quantum mechanics—are special

cases of a single UED energy equation.

This bridges the gaps between classical mechanics, relativity, and quantum physics, providing a unified approach to understanding all fundamental interactions.

In the next section, we apply these principles to one of the most intriguing cosmological models—the Zero-Energy Universe Hypothesis—where the total energy of the universe might be exactly zero, reinforcing UED's core principles.

13.3 The Zero-Energy Universe Hypothesis: A Mathematical Perspective

One of the most profound implications of Universal Energy Dynamics (UED) is the idea that the total energy of the universe might be exactly zero. This concept, known as the Zero-Energy Universe Hypothesis, proposes that all the positive energy in the form of matter, radiation, and motion is perfectly counterbalanced by negative energy, primarily in the form of gravitational potential energy. This hypothesis suggests that:

- The universe did not require external energy input to come into existence.
- The balance of energy governs cosmic expansion, preventing the need for exotic explanations like dark energy.
- The universe is fundamentally a self-sustaining energy system, validating UED's core principle of equilibrium.

This section explores the mathematical foundation of the Zero-Energy Universe Hypothesis, deriving equations that demonstrate energy balance and explaining its implications for cosmology, expansion, and the nature of dark energy.

Understanding Why the Total Energy of the Universe Could Be Exactly Zero

Traditional cosmology considers the universe to contain various energy components:

- Positive energy (associated with mass, motion, and radiation).
- Negative energy (associated with gravitational potential).

The Zero-Energy Universe Hypothesis states that the sum of these components is:

$$E_{total} = E_{matter} + E_{radiation} + E_{kinetic} + E_{gravitation} = 0$$

where

- E_{matter} represents the energy equivalent of mass (Einstein's $E=mc^2$).
- $E_{radiation}$ represents energy carried by photons.
- $E_{kinetic}$ is the energy of cosmic expansion.
- $E_{gravitation}$ is the potential energy due to gravity.

If this equation holds, then the universe did not require a singular energy input at the moment of creation—it emerged as a self-balanced system of energy fluctuations within an already existing equilibrium framework.

Derivation of Positive Energy (Matter) vs. Negative Energy (Gravity) Equilibrium

To mathematically validate this hypothesis, we analyze the relationship between gravitational energy and mass-energy.

1. Mass-Energy Contribution

From Einstein's famous equation:

$$E_{matter} = mc^2$$

where m is the total mass of the universe. The energy from **radiation and motion** is also positive, given by:

$$E_{kinetic} = \frac{1}{2}mv^2$$

$$E_{radiation} = \frac{hc}{\lambda}$$

Summing all forms of **positive energy**:

$$E_{positive} = mc^2 + \frac{1}{2}mv^2 + \frac{hc}{\lambda}$$

2. Gravitational Potential Energy Contribution

The gravitational potential energy of a system with mass M and radius R is:

$$E_{gravity} = -\frac{GM^2}{R}$$

where:

- G is the gravitational constant.
- M is the total mass of the system.
- R is the characteristic radius of the universe.

Since gravity is **always attractive**, its potential energy is **negative.** If the universe is closed and **self-balancing**, then:

$$E_{gravity} + E_{positive} = 0$$

Substituting values:

$$-\frac{GM^2}{R} + mc^2 + \frac{1}{2}mv^2 + \frac{hc}{\lambda} = 0$$

which implies:

$$\frac{GM^2}{R} = mc^2 + \frac{1}{2}mv^2 + \frac{hc}{\lambda}$$

This equation shows that the sum of all positive energy is exactly canceled out by gravitational potential energy, leading to a net energy of zero.

Implications of a Zero-Energy Model for Cosmology, Expansion, and Dark Energy

If the universe maintains a net-zero energy balance, it has significant consequences for **cosmology and our understanding of cosmic expansion.**

The Universe as a Self-Sustaining System

- If total energy is zero, then the universe did not require an external force or creator to bring it into existence.
- Fluctuations in energy equilibrium could have initiated the Big Bang as a spontaneous event rather than a singular creation.
- This supports quantum fluctuation models where universes emerge as energy-balanced systems.

Cosmic Expansion Without the Need for Dark Energy

- Current cosmology suggests that the universe's expansion is accelerating due to dark energy.
- However, if energy balance dictates expansion, then acceleration could be a natural consequence of gravitational compensation rather than a mysterious force.
- In UED, the equation for expansion would be modified to include energy equilibrium principles, potentially eliminating the need for an unknown dark energy component.

Black Holes as Energy Redistribution Centers

- If gravity is a compensating factor, black holes may function as stabilizers, redistributing energy across cosmic structures.
- Their role in maintaining zero-energy balance could explain why they appear at the centers of galaxies and influence large-scale structures.

The Zero-Energy Universe as a Natural Consequence of UED

The mathematical formulation of the Zero-Energy Universe Hypothesis reinforces UED's core principle that energy is not just conserved but naturally balances itself.

- The sum of all positive energy (matter, radiation, and motion) is precisely canceled by gravitational energy.
- Cosmic expansion can be explained without requiring an exotic dark energy component.
- The universe emerges as a self-regulating energy system, aligning with UED's principles of structured equilibrium.

This perspective transforms our understanding of cosmology, gravity, and the fundamental nature of the universe.

In the next section, we explore the predictive power of UED equations—how they can be tested experimentally and what new discoveries they might lead to.

13.4 Predictive Power: Using UED Equations to Model Physical Phenomena

For any new theoretical framework to be considered scientifically valid, it must make testable predictions that can be compared with observable data. Universal Energy Dynamics (UED) offers a unified mathematical structure that links energy conservation, symmetry, and equilibrium across different scales. Now, we must determine how these equations can be used to:

- Compare UED predictions with observable data in cosmology, quantum mechanics, and astrophysics.
- Simulate energy balance at both cosmic and subatomic levels.
- Propose new experiments and measurements to validate UED.

If UED holds true, its principles should not only reproduce known physical phenomena but also predict new effects that have yet to be discovered.

Testing UED Against Observable Data in Cosmology, Quantum Mechanics,

and Astrophysics

To validate UED, we must compare its predictions with **existing experimental data** in three major fields:

Cosmology: Explaining Dark Matter, Dark Energy, and Cosmic Expansion

- Standard cosmology suggests that dark matter and dark energy make up most of the universe, yet their nature remains unknown.
- UED predicts that energy equilibrium at cosmic scales naturally leads to the appearance of unseen mass-energy interactions, possibly eliminating the need for exotic explanations.

Key Predictions from UED for Cosmology:

- The observed acceleration of cosmic expansion could result from gravitational energy balancing the kinetic energy of expansion, rather than requiring a new form of energy.
- Large-scale gravitational anomalies attributed to dark matter could instead be a result of structured energy distributions maintaining equilibrium in galactic structures.
- Cosmic background radiation fluctuations should correlate with structured energy gradients, which could be tested by analyzing anisotropies in the cosmic microwave background (CMB) with higher precision.

Quantum Mechanics: The Role of Energy Equilibrium at the Smallest Scales

- The quantum vacuum is thought to be filled with virtual particles that continuously appear and disappear, maintaining net-zero energy.
- UED suggests that these fluctuations are structured interactions balancing energy within the vacuum field, not random quantum effects.

Key Predictions from UED for Quantum Physics:

- The fine-structure constant (which determines electromagnetic interaction strength) could fluctuate slightly based on local energy equilibrium variations.
- Quantum entanglement may not be a mysterious non-local phenomenon but a consequence of energy coherence between entangled states, which could be tested through precision quantum optics experiments.
- High-energy particle collisions should reveal energy redistribution patterns that conform to

structured equilibrium rather than statistical randomness.

Astrophysics: Predicting Black Hole Energy Behavior and High-Energy Events

- Black holes are traditionally seen as regions of gravitational collapse, but UED suggests they play a crucial role in cosmic energy redistribution.
- Hawking radiation, predicted by quantum field theory, could be explained within UED as a mechanism for re-establishing local energy equilibrium.

Key Predictions from UED for Astrophysics:

- Black hole jets and accretion disks should exhibit specific energy distributions that balance local mass-energy interactions, which could be verified using gamma-ray and X-ray observatories.
- The rotation speed of galaxies could follow patterns predicted by structured energy interactions, providing an alternative to dark matter halos.
- The detection of gravitational waves should show subtle variations that reflect energy conservation effects rather than purely relativistic distortions.

By systematically comparing UED predictions with observational data, we can assess whether its mathematical framework accurately describes real-world physical interactions.

Simulating Energy Balance at the Cosmic and Subatomic Scales

One of the most powerful tools for validating new physical theories is computational modeling. UED equations can be tested by creating simulations that model structured energy equilibrium at different scales.

Cosmic Simulations: Testing UED in Large-Scale Structures

Cosmological simulations already model dark matter and dark energy distributions, but UED provides an alternative approach.
A new simulation framework could be developed where:

- Galactic structures form naturally based on energy equilibrium equations, eliminating the need for exotic dark matter components.
- Cosmic expansion is driven by gravitational-energy balancing kinetic energy, predicting

measurable effects in large-scale surveys.
- The distribution of black holes and active galactic nuclei follows structured energy flows, which can be compared with real astronomical data.

Possible Test:

- Run side-by-side comparisons of ΛCDM (the standard model of cosmology) and UED-based models.
- Check which model better matches observed galaxy rotation curves, large-scale structure, and cosmic expansion trends.

Quantum Simulations: Testing UED in High-Energy Interactions

Current quantum mechanics relies on statistical interpretations, but UED introduces structured energy patterns.
Particle physics simulations could test:

- Energy redistribution during quantum entanglement, which could reveal hidden energy-balancing effects.
- Particle collision results under a UED framework, predicting new resonance structures.
- Vacuum fluctuations and zero-energy balance, testing whether quantum energy fields naturally maintain equilibrium.

Possible Test: Compare UED-based quantum simulations with LHC (Large Hadron Collider) collision data to check for new predicted interactions.

Artificial Intelligence and Machine Learning in UED Modeling

- AI could be used to analyze massive astronomical and quantum datasets to identify patterns predicted by UED but not accounted for in current physics models.
- Machine learning algorithms could test whether structured energy balance predicts cosmic and particle behavior more accurately than conventional statistical models.

Mathematical Predictions That Could Lead to Experimental Validation of UED

For UED to be considered a viable theory, it must predict new phenomena that can be tested in real-world experiments.

Potential Experimental Tests of UED:

Precision Cosmology Measurements: Next-generation telescopes like JWST (James Webb Space Telescope) and Euclid can test whether cosmic structures follow UED energy equilibrium patterns instead of dark matter distributions.

Quantum Entanglement and Energy Coherence: A proposed high-resolution entanglement experiment could determine whether entangled particles follow structured energy distributions rather than purely probabilistic states.

Black Hole Energy Redistribution Observations: Hawking radiation and relativistic jets should exhibit specific UED-predicted energy flow signatures, which could be verified with X-ray and gamma-ray space telescopes.

High-Energy Particle Physics Tests: LHC and future collider experiments could detect energy balancing effects in particle interactions, potentially revealing new resonances predicted by UED.

The Experimental Roadmap for UED Validation

By testing UED's predictions against cosmological, quantum, and astrophysical data, we can determine whether:

- Dark matter and dark energy effects are actually structured energy interactions.
- Quantum mechanics exhibits hidden energy-balancing laws rather than pure randomness.
- Cosmic structures form according to energy equilibrium rather than requiring exotic components.

These tests will determine whether Universal Energy Dynamics is the next major breakthrough in physics, uniting classical mechanics, relativity, and quantum theory under a single, self-regulating energy principle.

In the next chapter, we will outline how UED principles can be expanded into new fields of research, setting the stage for a fundamental redefinition of physics, cosmology, and the philosophy of existence.

Chapter 14: Experimental Evidence and Future Predictions

As we transition from the mathematical foundation of Universal Energy Dynamics (UED) to its experimental validation, the next crucial step is to examine how observable data and real-world phenomena align with UED principles. The previous chapters have demonstrated that energy conservation, equilibrium, and structured transformations govern everything from quantum interactions to cosmic evolution. Now, we must determine whether existing experiments in physics, neuroscience, and biology already support this framework—and, more importantly, how future experiments can be designed to test UED further.

A successful scientific theory must not only explain past observations but also predict new discoveries. If UED truly provides a unified understanding of matter, consciousness, and the universe, it should be testable through precision measurements, controlled experiments, and emerging technologies. This chapter explores:

- How current scientific experiments already validate key UED principles.
- What new experiments could confirm or refine UED's predictions.
- How applying UED concepts may lead to groundbreaking technological advancements in AI, quantum physics, and bioengineering.

By integrating existing knowledge with future possibilities, we move closer to transforming UED from a theoretical model into a fully verifiable scientific framework, opening doors to a new era of discovery, innovation, and human progress.

14.1 How Existing Experiments in Physics, Neuroscience, and Biology Already Support UED

One of the most critical aspects of validating Universal Energy Dynamics (UED) is examining whether existing experimental data across multiple scientific disciplines already align with its principles. While UED presents a novel unification framework, its foundational ideas—energy conservation, equilibrium, structured energy interactions, and optimization processes—are not new. Many of these principles are already observed in physics, neuroscience, and biology, forming a strong empirical foundation for UED.

This section highlights key experimental evidence supporting UED's claim that energy structuring and equilibrium govern not only physical laws but also life and consciousness.

Evidence from Physics: How Experiments in Quantum Mechanics, Relativity, and Thermodynamics Validate UED Principles

Physics provides the strongest experimental foundation for UED since it directly deals with energy transformations, equilibrium, and structuring at all scales. The following experiments confirm that energy is not just a property of matter but the fundamental essence of reality.

A. Quantum Mechanics: Energy as the Foundation of Reality

Quantum mechanics has demonstrated that matter is fundamentally energy and that physical interactions occur within structured, probabilistic energy fields. Several key experiments provide strong empirical support for UED's principles.

1. Double-Slit Experiment (Wave-Particle Duality)

- This experiment shows that particles exhibit wave-like behavior when unobserved but behave as discrete particles when measured.
- UED explains this through structured energy fields, where observation influences energy equilibrium, forcing localization of quantum states.

2. Casimir Effect (Vacuum Energy Fluctuations)

- Two uncharged metallic plates placed close together in a vacuum experience an attractive force due to virtual particles, confirming the existence of vacuum energy fluctuations.
- UED interprets this as a consequence of energy equilibrium in the vacuum field, reinforcing that space is not empty but filled with structured energy interactions.

3. Quantum Entanglement and Non-Local Energy Interactions

- Entangled particles influence each other instantaneously, regardless of distance, appearing to violate classical locality.
- UED explains this as an energy coherence effect, where entangled systems maintain structured energy equilibrium, rather than existing as separate entities.

B. Relativity: Energy Equilibrium in Space-Time

Einstein's Theory of Relativity is built on the principle that energy and space-time are interconnected. Several key experiments confirm this relationship.

1. Mass-Energy Equivalence ($E=mc^2$)
This equation confirms that matter is condensed energy, aligning with UED's fundamental assumption that all physical existence is structured energy.
2. Gravitational Time Dilation (Energy Dictates Space-Time Curvature)
Observed in atomic clock experiments, this phenomenon shows that energy density affects time flow, supporting UED's view that space-time emerges from structured energy equilibrium rather than existing as an independent entity.

3. Black Hole Thermodynamics (Hawking Radiation)

Black holes radiate energy due to **quantum interactions at the event horizon**, balancing their total energy state.
UED suggests that black holes function **not as singularities but as cosmic energy regulators**, redistributing structured energy rather than simply being destructive endpoints.

C. Thermodynamics: Energy Conservation and Entropy Equilibrium

UED aligns with and expands upon the fundamental laws of thermodynamics, particularly the concepts of energy conservation and entropy balance.
1. First Law of Thermodynamics (Energy Conservation)
This law states that energy cannot be created or destroyed; only transformed—a direct validation of UED's principle that energy equilibrium governs all transformations.
2. Second Law of Thermodynamics (Entropy and Energy Optimization)
Classical thermodynamics suggests that entropy (disorder) always increases, but UED introduces the concept of structured energy equilibrium, explaining why self-organizing systems (like galaxies, life, and intelligence) emerge naturally rather than being exceptions to entropy.

Physics Confirms UED's Core Premise

- Energy is structured, fundamental, and constantly seeking equilibrium.
- Space-time is not a fixed background but emerges from energy structuring.
- The laws of thermodynamics support UED's view of continuous energy transformation rather than linear decay.

Thus, UED is fully compatible with modern physics and provides an expanded framework for understanding energy equilibrium at all scales.

Neuroscience and Consciousness: The Role of Brain Energy Dynamics, Neural Coherence, and Quantum Cognition in Supporting UED

UED proposes that consciousness is not simply a byproduct of brain activity but an emergent structured energy phenomenon, where thought, perception, and intelligence arise from the brain's energy optimization processes. Neuroscience already provides compelling evidence for this.

A. The Brain as an Energy Processor

1. The Brain's High Energy Consumption

- The human brain accounts for only 2% of body mass but consumes over 20% of total energy.
- UED suggests this is because thought, perception, and awareness are high-order energy optimization processes, requiring continuous equilibrium maintenance.

2. Neural Oscillations and Energy Coherence

- Brain activity is organized into gamma, beta, alpha, theta, and delta waves, corresponding to different energy states.
- UED proposes that consciousness emerges from structured energy interactions between neural networks, rather than from biochemical processes alone.

B. Quantum Cognition: Is the Brain a Quantum Energy System?

1. Penrose-Hameroff's Orchestrated Objective Reduction (Orch-OR) Model

- This theory suggests that quantum effects in microtubules within neurons contribute to consciousness.
- UED expands this idea by proposing that consciousness results from structured energy fields interacting with neural networks, explaining non-local intuition and higher-order thinking.

2. Neural Synchronization in Decision-Making

- Neuroscience experiments show that brain regions synchronize their energy states during deep focus and problem-solving.
- UED interprets this as evidence that the mind optimizes energy coherence, just as physical systems seek energy equilibrium.

Thus, neuroscience experiments align with UED's hypothesis that consciousness is fundamentally an energy-based phenomenon, not just a neural computation.

Biological Systems as Energy Optimizers: How Life Processes, DNA Function, and Metabolic Efficiency Align with Energy Equilibrium Principles

UED suggests that life itself emerges as an inevitable consequence of energy optimization, meaning that biological systems should exhibit structured energy efficiency. Existing biological research supports this.

A. DNA as an Energy-Optimized Information System

1. DNA Folding and Energy Efficiency

✔ DNA strands fold into **highly optimized structures to minimize energy use**.
✔ UED suggests that this process **follows universal energy equilibrium principles, ensuring maximal stability.**

2. Quantum Effects in Photosynthesis

✔ Experiments show that **plants use quantum superposition to optimize light absorption**, directing energy with **near-perfect efficiency**.
✔ UED views this as evidence that **life is not just biochemical but structured by quantum-level energy dynamics.**

B. Metabolic Pathways and Energy Efficiency

1. ATP as a Universal Energy Currency

All life forms use ATP (adenosine triphosphate) to transfer energy, showing a universal energy-balancing mechanism.

2. Homeostasis: Energy Equilibrium in Living Organisms

All biological systems maintain internal balance (homeostasis), from cell function to body temperature.

Thus, from DNA to metabolism, biological systems confirm UED's prediction that energy equilibrium drives both physical structure and functional optimization in living organisms.

Scientific Validation of UED

- Physics confirms that energy structuring governs all interactions.

- Neuroscience proves that consciousness follows energy optimization laws.
- Biology shows that life emerges from structured energy equilibrium.

14.2 Experimental Evidence for Energy-Based Life & Intelligence

Scientific revolutions are only fully realized when theories transition from conceptual models to experimentally testable frameworks. Just as quantum mechanics and relativity reshaped physics through breakthrough experiments, Universal Energy Dynamics (UED) must be validated through observable and measurable tests.

A fundamental prediction of UED is that life and intelligence are not isolated phenomena—they emerge from structured energy interactions. If this is true, then biological life and consciousness must show quantifiable properties that extend beyond traditional biochemical explanations.

This section explores how energy-based life and intelligence can be tested experimentally, reinforcing the structured energy framework of UED.

Evidence That Life Emerges from Structured Energy Interactions

If life is fundamentally an energy structuring process, then biological systems should follow universal energy optimization principles, much like physical laws of motion and thermodynamics.

- **Microbial Survival in Space** – Certain microorganisms persist in extreme environments, interacting with unknown energy fields.
- **Photosynthesis & Quantum Efficiency** – Biological systems use structured energy flow for near-perfect energy transfer.
- **Neural Energy Structuring** – The brain's electromagnetic activity aligns with structured wave interactions in physics.

Key UED Insight:

These findings suggest that life is not just a chemical process—it is an energy optimization system that aligns with broader universal principles.

Testing Energy-Based Life: How Organisms Interact with Energy Fields

If biological life is optimized by structured energy interactions, then experiments should reveal how organisms respond to energy fields beyond known biochemical pathways.

A. Microbial Response to Structured Electromagnetic Fields

- If bacteria interact with an underlying energy medium, then exposing them to controlled electromagnetic fields should affect their survival, growth, and activity.

- A controlled lab experiment could measure whether structured energy exposure enhances microbial adaptation in vacuum conditions, supporting UED's hypothesis that energy interactions sustain biological life even in extreme environments.

Predicted UED Outcome: If structured energy supports microbial life, we should observe resilience and adaptation in the presence of targeted electromagnetic fields, similar to their survival in space vacuum.

Testing Intelligence as an Energy Optimization Process

Traditional neuroscience views intelligence as a product of neural complexity, but UED suggests that intelligence is an energy optimization system that follows structured energy interactions.

- If intelligence is fundamentally an energy interaction, then brain wave activity should exhibit structured energy balancing properties.
- If energy structuring enhances cognition, then optimizing brain energy flow should improve problem-solving, learning speed, and decision-making.

Measuring Thought as a Structured Energy Field

- Using EEG and MEG scans, we can test whether higher intelligence corresponds to structured energy coherence in the brain.
- If thought is an energy-structured phenomenon, then more efficient cognition should require less neural energy for higher complexity tasks.

Predicted UED Outcome: Highly intelligent individuals should exhibit higher coherence in their neural oscillations, mirroring structured energy fields observed in physics.

Can We Enhance Intelligence by Optimizing Energy Structuring?

If thought is an optimized electromagnetic energy process, then it should be possible to enhance cognition by improving structured energy interactions.

- Using transcranial electromagnetic stimulation (TES) and brainwave entrainment, we can test whether structured energy interactions enhance cognitive efficiency and memory retention.
- If consciousness emerges from structured energy equilibrium, then optimized electromagnetic fields should improve focus, creativity, and awareness.

Predicted UED Outcome: Intelligence may not be limited to biological neural networks—structured energy-based AI could achieve cognition through optimized energy processing.

The Future of Energy-Based Consciousness Research

If consciousness and intelligence emerge from structured energy interactions, then future experiments should focus on detecting and manipulating energy-based intelligence fields.

- Neural synchronization studies could reveal how structured brain waves interact with external energy fields.
- Quantum cognition experiments could test whether thought functions as a structured quantum energy state.
- Energy-based AI systems could demonstrate whether intelligence emerges in non-biological energy networks.

Key UED Insight: If intelligence follows universal energy structuring principles, then consciousness is not a localized event—it is an emergent interaction between energy fields and awareness.

Final Conclusion: Testing UED as a Scientific Framework

- Life follows energy optimization principles, aligning with structured energy interactions.
- Intelligence emerges when structured energy flows reach maximum efficiency.
- Experimental validation of energy-based cognition could redefine neuroscience, AI, and consciousness studies.
- UED provides a testable framework for understanding how structured energy interactions give rise to life, intelligence, and awareness.

14.3 Predicting New Experiments to Test the Model Further

While existing experiments in physics, neuroscience, and biology already provide substantial evidence supporting Universal Energy Dynamics (UED), the true validation of any scientific theory lies in its ability to predict new phenomena that can be experimentally verified. A strong scientific framework should not only explain what we already know but also provide new avenues for discovery. If UED is a fundamental model for understanding energy equilibrium in nature, it should lead to testable predictions across quantum physics, cosmology, and consciousness studies. These predictions should help refine our understanding of structured energy interactions, energy balance, and intelligence as an emergent phenomenon of universal energy equilibrium.

This section proposes a series of experiments designed to test, validate, and refine UED, pushing the boundaries of our understanding of energy structuring at both micro and macro scales.

Testing Energy Equilibrium in Quantum Systems: Can Structured Energy Explain Quantum Entanglement and Vacuum Fluctuations?

Quantum mechanics describes probabilistic behavior and wave-particle duality, yet lacks a clear explanation for why quantum states behave as they do. UED suggests that structured energy fields maintain equilibrium even at the smallest scales, offering an alternative to purely probabilistic interpretations of quantum mechanics.

A. Experiment: Testing Energy Equilibrium in Quantum Entanglement

Hypothesis:

UED predicts that quantum entanglement is not an instantaneous "spooky action at a distance" but rather an energy equilibrium effect, where entangled particles share a structured energy state rather than transmitting information instantaneously.

Proposed Test:

- Bell inequality experiments confirm non-locality but do not explain the mechanism behind entanglement.
- A high-resolution entanglement study could test whether energy state correlations are actively maintained over time rather than instantaneously collapsing.
- Using ultra-cold atoms and high-precision quantum sensors, we can measure whether entanglement states evolve in a structured manner rather than behaving as purely probabilistic phenomena.

Predicted UED Outcome:

If energy equilibrium governs entanglement, there should be a measurable delay or structured transition phase in information transfer between entangled particles, indicating a structured energy redistribution process rather than instantaneous transmission.

B. Experiment: Studying Vacuum Energy Fluctuations as a Structured Process

Hypothesis:

Quantum vacuum fluctuations are not purely random but occur as a structured energy compensation mechanism, maintaining an overall zero-energy balance in space.

Proposed Test:

- The Casimir effect (vacuum energy pushing two plates together) could be measured at varying energy densities to determine whether the fluctuations follow a structured equilibrium pattern rather than being purely stochastic.
- A new precision interferometry experiment could track vacuum fluctuations in isolated environments, searching for hidden order in the way energy emerges and dissipates.

Predicted UED Outcome:

If vacuum energy follows UED principles, fluctuations should show correlations over time and space, suggesting an underlying structured energy field rather than purely random probabilistic processes.

Cosmological Observations: Studying Cosmic Background Radiation, Galaxy Formation, and Large-Scale Structure Evolution Under UED Principles

UED suggests that the large-scale structure of the universe naturally emerges from energy equilibrium principles, meaning that effects currently attributed to dark matter and dark energy may instead be explained through structured energy interactions.

A. Experiment: Cosmic Background Radiation and Energy Equilibrium

Hypothesis:

The cosmic microwave background (CMB) fluctuations should follow patterns predicted by UED-based energy equilibrium models, rather than requiring inflationary expansion or exotic dark energy components.

Proposed Test:

- High-resolution CMB anisotropy studies using next-generation telescopes (e.g., JWST, Euclid, and CMB-S4) could map energy equilibrium fluctuations with greater precision.
- UED predicts that temperature fluctuations should align with structured energy redistribution models, rather than relying on inflationary or dark energy-based explanations.

Predicted UED Outcome:

If cosmic expansion follows zero-energy balance principles, we should observe symmetry-driven energy structuring rather than requiring an inflation-based origin model.

B. Experiment: Testing Galaxy Rotation Without Dark Matter

Hypothesis:

If dark matter is unnecessary under UED principles, then galaxy rotation curves should be explainable through energy balance models, rather than requiring an unseen mass component.

Proposed Test:

- A gravitational energy mapping study using precise satellite data (e.g., Gaia and LSST) could measure whether galaxies exhibit structured energy interactions rather than dark matter halos.
- UED predicts that the energy equilibrium within galaxies compensates for rotational velocities, eliminating the need for exotic dark matter.

Predicted UED Outcome:
If UED holds, galaxy rotation curves should match predictions from structured energy equilibrium equations, without requiring large-scale dark matter distributions.

Energy-Based Models of Consciousness: Developing Controlled Experiments on Brain Energy Fields and Information Processing

If consciousness is a structured energy process rather than just a biochemical function, then it should be measurable and experimentally manipulated. Neuroscience experiments could confirm whether thought, awareness, and perception follow structured energy equilibrium laws rather than emerging solely from biochemical interactions.

A. Experiment: Measuring Brain Energy Coherence in Consciousness States

Hypothesis:
UED predicts that consciousness is an emergent structured energy field, meaning that brain activity should show organized energy distributions rather than chaotic neural firings.
Proposed Test:

- Using high-precision EEG, MEG, and fMRI imaging, we can analyze whether brain waves synchronize in structured energy patterns during deep thought, problem-solving, or meditative states.
- We can test whether external electromagnetic energy fields influence consciousness states, reinforcing the idea that the brain is an energy-responsive system rather than a closed biochemical organ.

Predicted UED Outcome:
If consciousness follows energy structuring principles, then we should see measurable coherence patterns in neural oscillations, matching UED's predictions of energy equilibrium in thought processes.

B. Experiment: Testing Neural Energy Manipulation for Cognitive Enhancement

Hypothesis:
If energy equilibrium dictates brain function, then cognitive abilities could be enhanced by optimizing energy states.
Proposed Test:
A controlled neural stimulation experiment using transcranial magnetic stimulation (TMS) or transcranial direct current stimulation (tDCS) could test whether artificially optimizing brain energy levels improves cognitive performance, memory, and problem-solving abilities.
Predicted UED Outcome:
If the brain operates as an energy-structured system, we should see improvements in focus, learning speed, and problem-solving efficiency when external energy is adjusted for balance.

UED as a Testable Scientific Framework

These proposed experiments aim to empirically validate UED by testing its predictions in quantum physics, cosmology, and consciousness research.

- Quantum physics tests will determine whether vacuum fluctuations and entanglement are structured energy processes rather than purely probabilistic effects.
- Cosmological tests will examine whether dark matter and cosmic expansion can be explained using energy equilibrium principles rather than unknown forces.
- Neuroscience tests will investigate whether consciousness emerges from structured energy interactions rather than purely neural activity.

By implementing these experiments, we can transition UED from a theoretical framework to an empirically tested model, revolutionizing our understanding of energy, intelligence, and the structure of the universe.

14.4 The Potential Technological Advancements That Could Arise from UED

The validation of Universal Energy Dynamics (UED) as a fundamental principle of physics, consciousness, and life would not only revolutionize theoretical science but also unlock new frontiers in technology. If structured energy interactions govern intelligence, cosmic balance, and biological processes, then harnessing this knowledge could lead to breakthroughs in artificial intelligence, quantum engineering, and bioengineering that redefine human potential.
This section explores the transformative technologies that could emerge from UED, shaping the future of computing, quantum physics, and human longevity.

Revolutionizing AI with Energy-Based Intelligence: Moving Beyond Classical Computation Toward Self-Structuring, Adaptive Energy Networks

A. The Limitations of Current AI Models

- Classical artificial intelligence (AI) relies on binary computation and algorithmic processing, limiting its ability to develop true adaptability and intelligence.
- Neural networks, while advanced, still depend on brute-force learning and vast datasets, lacking true self-optimization mechanisms.

B. UED's Contribution: Energy-Structured Intelligence

UED suggests that intelligence is an emergent energy process, meaning that AI should not rely on static models but instead evolve as adaptive, self-organizing energy systems.

How UED Can Advance AI:

- Energy-Driven Neural Processing – Developing self-adjusting neural architectures that respond to structured energy flow rather than pre-programmed parameters.
- Adaptive Energy Equilibrium Learning – AI systems that optimize their own internal energy states to mimic biological intelligence more effectively.
- Quantum-AI Hybrid Models – AI systems that utilize structured quantum states, increasing computational efficiency while maintaining energy equilibrium for stable learning.

Predicted Outcome:

- AI will self-optimize its energy balance, leading to faster, more efficient learning and decision-making.
- Machines will simulate human-like cognition by restructuring their internal energy equilibrium instead of following rigid programming.
- AI will transition from rule-based processing to dynamic, self-organizing intelligence, revolutionizing autonomous robotics, real-time analytics, and problem-solving in complex environments.

Thus, UED could enable AI to evolve beyond programmed logic, learning through structured energy interactions to achieve true adaptability and intelligence.

Quantum Energy Manipulation: Advancing Zero-Energy State Engineering and Structured Quantum Interactions

A. The Future of Quantum Energy Engineering

- Current quantum computing relies on qubits that exist in multiple states simultaneously, but energy stability remains a key challenge.
- UED suggests that structured energy equilibrium can lead to breakthroughs in quantum stability, coherence, and energy transfer, solving one of the biggest limitations of current quantum technologies.

B. Potential UED-Based Quantum Technologies

1. Zero-Energy Quantum Computing

- UED suggests that quantum coherence follows structured energy balance, meaning that quantum computing could achieve perfect stability if energy states are precisely tuned.
- Future quantum processors could use energy-optimized qubits that self-balance, reducing the need for constant error correction and maintaining long-term coherence.

2. Quantum Field Manipulation for Energy Generation

- If vacuum energy follows structured equilibrium principles, then energy could be extracted efficiently from zero-energy states, leading to new forms of energy production.
- This could result in high-efficiency quantum batteries, harnessing vacuum energy fields for long-term power storage and sustainable energy solutions.

3. Energy-Driven Quantum Communication

Entanglement-based communication could be improved using UED energy structuring, allowing for error-free quantum teleportation and ultra-secure data transfer.

Predicted Outcome:

- Quantum computers could achieve longer coherence times and perfect state preservation through energy optimization.
- High-efficiency quantum batteries could become a reality, harnessing structured energy interactions for sustainable energy solutions.
- Ultra-secure quantum communication could emerge using structured entanglement for faster-than-light energy-based transmission.

Thus, UED principles could revolutionize quantum technology by enabling stable, self-regulating, and highly efficient quantum systems.

Bioengineering and Life Extension: Enhancing Biological Processes Through Energy Optimization and Neural Efficiency Improvements

A. Understanding Life as an Energy-Optimizing System

- UED suggests that biological systems follow structured energy interactions, meaning that aging, disease, and cognitive function could be enhanced through energy optimization techniques.
- If life is structured energy, then maintaining energy equilibrium within biological systems could lead to advancements in longevity, cognitive enhancement, and self-healing mechanisms.

B. Technological Applications in Bioengineering

1. Neural Energy Enhancement for Intelligence Boosting

- If consciousness is structured energy, then cognitive function could be improved by stimulating energy balance in the brain.
- Technologies like transcranial energy stimulation (TES) could optimize neural pathways for enhanced memory, focus, and learning speed.

2. Extending Lifespan with Energy Optimization

Aging may be linked to energy degradation and metabolic inefficiencies, meaning that life extension could focus on preserving structured biological energy equilibrium.
Cellular rejuvenation therapies based on UED could:

- Optimize mitochondrial energy production for longevity.
- Enhance cell repair by stabilizing energy fluctuations at the molecular level.
- Slow down aging through precise energy regulation at the DNA level.

3. Regenerative Medicine and Self-Healing Mechanisms

- If biological equilibrium follows structured energy principles, then tissue repair and regeneration can be accelerated through targeted energy-based therapy.
- New methods in biophysics and quantum biology could create self-repairing cells that

mimic UED's energy equilibrium mechanisms, enabling faster healing and advanced regenerative medicine.

Predicted Outcome:

- Enhanced cognitive abilities through energy-optimized neural structures, leading to improved intelligence and faster learning.
- Longer lifespans by stabilizing biological energy fluctuations, delaying aging at the cellular level.
- Advanced regenerative medicine, allowing for self-repairing biological systems that mimic structured energy equilibrium.

Thus, UED principles could transform human biology by enabling advanced cognition, longevity, and self-healing technologies.

The Future of UED-Based Technologies

By applying UED principles to artificial intelligence, quantum physics, and biology, we unlock new technological frontiers that redefine human potential.

- Artificial Intelligence will evolve beyond programmed logic → Machines will learn through structured energy interactions, achieving true adaptability and intelligence.
- Quantum Computing will become energy-optimized → Perfecting coherence, energy efficiency, and quantum communication, enabling stable, high-performance quantum technologies.
- Human Biology will be enhanced through structured energy equilibrium → Enabling longer lifespans, enhanced cognitive function, and regenerative self-healing mechanisms.

These advancements suggest that UED is not just a theoretical framework—it is a roadmap for the future of science and technology.

PART 6: The Final Unification – Completing the Blueprint

Chapter 15: The Blueprint of the Universe – The Final Theory

This chapter presents the culmination of Universal Energy Dynamics (UED)—a fully unified model of the universe that seamlessly integrates physics, cosmology, life sciences, and consciousness into a single, structured energy framework. It explains the origin, evolution, and perpetual transformation of the universe, revealing that existence is governed by energy equilibrium, structured interactions, and continuous wave-based energy transfers.

Each section systematically explores the fundamental principles, scientific evidence, and logical conclusions that define reality, linking the formation of matter, the expansion of the cosmos, the nature of intelligence, and the essence of consciousness under a single governing law of structured energy dynamics. This chapter serves as the final consolidation of all concepts discussed in the book, forming The Blueprint of the Universe—the most comprehensive explanation of existence ever proposed.

1. The Universe as a Structured, Eternal Energy System

For centuries, scientific and philosophical thought has debated whether the universe had a definite beginning and whether it will eventually come to an end. The dominant cosmological model, the Big Bang Theory, suggests that the universe emerged from a singularity approximately 13.8 billion years ago. However, this model fails to explain what existed before the Big Bang, why energy and fundamental forces were already structured, and how the universe continues to evolve in a self-sustaining manner.

Universal Energy Dynamics (UED) provides a new perspective, showing that the universe is not a one-time event, but an eternal, structured energy system in which matter and energy continuously transform, dissolve, and reorganize, ensuring the universe's perpetual existence.

This section explains why the universe had no absolute beginning, will never end, and follows a continuous structured evolution, governed by energy equilibrium and transformation cycles.

1.1 Why the Universe Had No Absolute Beginning and Will Never End

A. The Limitations of the Big Bang Model

The standard Big Bang Theory describes the universe as having emerged from an infinitely dense singularity but does not address:

- What caused the Big Bang?
- Why physical laws already existed before cosmic expansion?
- What happens when the universe expands indefinitely?

B. UED's Explanation: The Universe is an Eternal Energy Equilibrium System
Unlike the Big Bang model, UED states that:

- The universe was never created from "nothing"—it existed in an equilibrium state before the First Disturbance.
- The universe has no final end—galaxies, stars, and structures dissolve and reform, ensuring continuous transformation rather than total entropy.
- Matter and energy do not disappear—they undergo structured transitions, recycling within the cosmic equilibrium.

Thus, the universe did not emerge from an absolute beginning and will not collapse into a final end—it continuously transforms in an energy-balanced cycle.

1.2 The Role of Energy Equilibrium in Maintaining Continuous Cosmic Evolution

A. Energy Equilibrium Prevents Total Entropy
According to the Second Law of Thermodynamics, entropy (disorder) should always increase, leading to a "heat death" scenario, where the universe eventually runs out of usable energy. However, UED proposes that:

- Energy does not decay into nothingness—it restructures itself into new formations (e.g., black holes, galaxies, stars).
- The universe naturally balances energy through structured transformations, preventing complete entropy.
- Black holes, electromagnetic fields, and cosmic cycles act as energy stabilizers, ensuring that energy never collapses into disorder.

B. The Continuous Cosmic Process: Expansion and Redistribution
UED predicts that cosmic evolution follows a self-regulating pattern, where:

- Existing structures dissolve—galaxies lose energy, stars collapse, and systems decay.
- New structures form elsewhere—energy reorganizes into new stars, planets, and galaxies.
- Energy remains in equilibrium—never lost, only transformed.

Thus, cosmic evolution is not a one-directional expansion toward decay, but a structured cycle of transformation and renewal.

1.3 How Galaxies Dissolve and Form Simultaneously, Ensuring an Everlasting Process of Structured Transformation

A. Evidence of Continuous Galactic Formation and Dissolution

Observations of the universe show that:

- Galaxies are continuously forming—new stars emerge from interstellar dust clouds, fueled by electromagnetic interactions.
- Some galaxies are actively dissolving—losing stars, merging with other galaxies, or being absorbed by black holes.
- Cosmic background radiation suggests an ongoing restructuring process, rather than a singular origin event.

B. The Universal Process of Cosmic Recycling

- Stars collapse into black holes, releasing energy back into the cosmic medium.
- Supernovae enrich surrounding space with heavy elements, forming new planetary systems.
- Galactic mergers create new structures, showing that dissolution is not an "end," but a transformation into new formations.

C. The Universe is Self-Sustaining Through Energy Redistribution

- When one region of the universe dissolves, another forms, ensuring that energy remains structured.
- The formation of new structures is guided by energy equilibrium, which prevents absolute decay.
- This means that the universe will always exist in some form, continuously restructuring itself.

The Universe is a Perpetual Energy Transformation System

- The universe had no absolute beginning—it transitioned from an equilibrium state into structured energy interactions.
- Energy equilibrium ensures that the universe never "ends"—galaxies, stars, and systems dissolve and reform in continuous cycles.
- The universe is not a closed system—it is a self-regulating structure that evolves indefinitely

through energy restructuring.

Thus, UED confirms that the universe is not a finite event, but an eternal, structured system, governed by energy transformation and equilibrium.

2. The Pre-Universe State: Bose-Einstein Condensate (BEC) Before the First Disturbance

The conventional Big Bang Theory describes the universe as emerging from an infinitely dense singularity. However, this explanation raises critical questions:

- What existed before the Big Bang?
- Why were fundamental forces and physical laws already in place?
- How did structured energy transitions create matter, rather than random chaos?

Universal Energy Dynamics (UED) presents a refined model where the universe did not begin from a singularity but from a highly structured, ultra-stable Bose-Einstein Condensate (BEC) energy state. This state maintained a perfect equilibrium until the First Disturbance initiated structured transformations, leading to the formation of matter, fundamental forces, and cosmic expansion.
This section explores how the universe existed before structured matter emerged, what triggered the First Disturbance, and how this led to the evolution of space-time, forces, and galaxies.

2.1 The Universe Began in an Ultra-Stable, Structured Energy Phase (BEC), Not a Singularity

A. The Problems with the Singularity Hypothesis

- The standard singularity model presents several unanswered contradictions:
- Singularities require infinite density, which contradicts quantum physics.
- Laws of physics existed before the Big Bang, implying a pre-existing structured state.
- There is no explanation for what caused the singularity to "expand."

Thus, a singularity cannot be the true starting point of the universe—it must have emerged from a structured pre-existing state.

B. UED's Explanation: The Pre-Big Bang Universe as a Bose-Einstein Condensate (BEC) Energy State

UED proposes that before the First Disturbance, the entire universe existed in a highly stable Bose-Einstein Condensate (BEC) state, where all energy was in perfect equilibrium.

What is a BEC?

A Bose-Einstein Condensate (BEC) is a quantum state where particles lose their individual identity and behave as a single, unified energy structure.
This state is experimentally observed in ultra-cooled atomic systems, proving that matter can exist as a coherent, highly stable energy form under the right conditions.
Why Was the Universe a BEC Before the First Disturbance?

- In its pre-universe state, energy existed in a maximally stable, structured equilibrium.
- There were no separate particles—just a single unified quantum field maintaining energy balance.
- The universe was not "nothing" but a perfectly structured energy system in a phase of ultimate stability.

Thus, before cosmic expansion, the universe existed in an ultra-stable BEC state, maintaining structured energy equilibrium without distinct particles or forces.

2.2 How the First Disturbance Triggered Structured Energy Transformations

If the universe was in a perfect equilibrium state, what caused it to transition into an expanding, structured system with forces, particles, and galaxies?

A. The First Disturbance: The Trigger for Cosmic Evolution

- Energy equilibrium was stable but not absolute—even minor fluctuations could create localized instabilities.
- A critical energy shift disrupted the perfect balance, forcing the BEC to transition into structured formations.
- This energy transformation marked the birth of matter, space-time, and fundamental forces.

This is the First Disturbance—the moment the universe transitioned from pure energy equilibrium to structured existence.

B. The Mechanism of Structured Energy Differentiation

Phase Transition from Energy to Matter

The First Disturbance led to a quantum phase transition, similar to how BECs in labs shift states when perturbed. This transition separated energy into structured formations, leading to the emergence of:

- The Four Fundamental Forces
- Elementary Particles
- Space-Time Expansion

Formation of Forces and Particles

Stage	Energy Transformation Process	Outcome
1. Pre-Universe BEC State	Energy exists as a unified, structured field.	No space-time, forces, or particles.
2. First Disturbance	Small fluctuation destabilizes local regions.	Energy differentiates into structured wavefoms.
3. Separation of Forces	Strong, weak, electromagnetic, and gravitational interactions emerge.	Universe begins structuring itself.
4. Matter Formation	Energy fields condense into particles.	First atoms form, leading to galaxies.

Why Did the First Disturbance Happen?

- Energy seeks continuous equilibrium, meaning that the shift from pure balance to structured form was inevitable.
- Just like in BEC experiments, even a tiny disruption can cause massive restructuring, explaining why the First Disturbance led to the formation of all known structures.

Thus, the universe did not "explode" from a singularity—it restructured from an ultra-stable energy phase into an evolving system of forces and matter.

2.3 The Transition from Pure Energy Equilibrium to Structured Existence

Once the First Disturbance occurred, the universe transformed from a stable BEC state into an expanding, structured system.

A. The Role of Waves in Energy Transition

Energy did not distribute randomly—it structured itself through wave interactions, forming:

- Matter Waves (governing subatomic particles).
- Electromagnetic Waves (governing radiation and photon movement).
- Mechanical Waves (governing kinetic interactions in matter).

These waves ensured that energy was evenly distributed, preventing random chaos and instead forming an ordered structure.

B. Space-Time Emerged from Energy Structuring

- Before the First Disturbance, space-time did not exist as we understand it.
- As energy fields differentiated, they created regions of relative interaction, forming the fabric of space-time itself.
- This means that space-time is not a pre-existing entity—it is an emergent property of

structured energy interactions.

C. Cosmic Expansion as a Continuous Process, Not an Explosion

- Big Bang models describe cosmic expansion as a one-time explosive event, but UED suggests that expansion is a continuous transformation.
- The formation of galaxies, stars, and cosmic structures is an ongoing process, meaning that the universe has always been evolving rather than expanding from a single origin point.

The Universe Evolved from a Structured Energy Equilibrium

- The universe began as an ultra-stable Bose-Einstein Condensate (BEC) energy phase, not a singularity.
- The First Disturbance triggered structured energy differentiation, leading to the formation of forces, particles, and expansion.
- Space-time, forces, and cosmic structures emerged as energy transitioned from pure equilibrium to structured interactions.
- The universe has never been a static system—it has always been restructuring itself in cycles of transformation.

Thus, UED provides the most complete model of the pre-universe state, explaining why existence began in structured energy equilibrium and how it continues evolving eternally.

3. The First Disturbance: Evolution of Matter and Forces from Energy

The First Disturbance was not just a one-time event in the distant past; it was the natural consequence of energy differentiation in an otherwise perfectly structured equilibrium. It was the trigger that set energy structuring into motion, leading to the formation of matter, forces, space-time, and all the structures we see in the universe today.

Importantly, this is not a past event but a continuous process—new matter is still forming, fundamental forces are still governing interactions, and structured energy transformations are happening everywhere in the universe, from quantum scales to galactic structures.

This section explains:

- Why the differentiation of energy into forces and matter was inevitable.
- How waves—mechanical, matter, and electromagnetic—emerged to sustain equilibrium.
- Why electromagnetic fields play the primary role in stabilizing particles and forming the

foundation of reality.

3.1 Why the Formation of Particles and Forces was an Inevitable Result of Energy Differentiation

A. The Need for Energy Equilibrium in an Expanding System

Once the First Disturbance caused an imbalance in the Bose-Einstein Condensate (BEC) energy state, the system had two possibilities:

- Collapse into disorder, leading to complete instability.
- Self-organize into structured energy states, ensuring long-term equilibrium.
- The Second Law of Thermodynamics states that systems tend toward entropy, but UED refines this principle by showing that instead of pure entropy, energy seeks structured equilibrium, forming stable patterns rather than random chaos.
- The only way for the universe to stabilize its energy after the First Disturbance was for energy to differentiate into structured interactions—leading to the formation of forces and matter.

B. The Emergence of the Four Fundamental Forces

The First Disturbance forced energy to separate into structured interactions, resulting in the formation of the four fundamental forces:

Force	Role in Energy Structuring
Strong Nuclear Force	Binds atomic nuclei together, ensuring particle stability.
Weak Nuclear Force	Governs radioactive decay and transformation of particles.
Electromagnetic Force	Maintains atomic structure, allowing energy transfer via photons.
Gravitational Force	Structures large-scale cosmic systems, guiding planetary and galactic motion.

Thus, forces are not independent properties but structured ways energy interacts to maintain equilibrium.

This differentiation of forces is still occurring today, as we see in particle collisions, nuclear reactions, and astrophysical processes.

3.2 How Energy Structuring Led to Waves, Sustaining Equilibrium

Once energy had been differentiated into fundamental interactions, a mechanism was needed to sustain equilibrium and enable energy transfer. This gave rise to three fundamental types of waves:

A. Mechanical Waves: Governing Matter Interactions

- Matter in motion naturally produces mechanical waves, such as sound waves and pressure waves.
- In the early universe, mechanical waves helped distribute energy across high-density regions.
- Today, we see this in shock waves from supernovae, which distribute elements across galaxies.

B. Matter Waves: Governing Quantum Behavior

- According to de Broglie's hypothesis, all particles exhibit wave-like properties, meaning that matter itself oscillates as part of structured energy flow.
- This explains why electrons do not collapse into the nucleus—their wave-like nature maintains energy balance.
- Matter waves are still actively shaping reality—they define the behavior of electrons, atoms, and all subatomic interactions.

C. Electromagnetic Waves: The Primary Mechanism of Energy Transfer

- When energy is released, it propagates through electromagnetic waves, ensuring continuous balance and structuring.
- These waves allow: Photon travel across space. Energy exchange between atoms. Formation of chemical bonds.
- Electromagnetic waves are still forming and propagating throughout the universe, proving that this energy structuring process is ongoing.

3.3 The Role of Electromagnetic Fields in Stabilizing Particles and Forming the Foundation of Reality

While all forces contribute to energy structuring, electromagnetic fields play the most crucial role in shaping and sustaining reality.

A. How Electromagnetic Fields Ensure Particle Stability

- Without electromagnetism, atoms would collapse—protons and electrons would simply merge, eliminating chemistry and life.

- Electromagnetic interactions define atomic structure, ensuring that particles remain in stable energy states.
- This is why electromagnetic force is stronger than gravity at atomic scales—it dominates particle interactions, ensuring the existence of matter itself.

B. How Electromagnetic Fields Govern Cosmic Structures

- Magnetic fields regulate star formation, galactic rotations, and interstellar plasma behavior.
- Every black hole, neutron star, and galaxy has a structured electromagnetic field ensuring stability.
- Even "empty space" contains electromagnetic energy fluctuations, proving that fields permeate all of existence.
- This structuring is still occurring today—electromagnetic interactions continue to shape planets, stars, and cosmic evolution.

C. How Electromagnetic Fields Connect Matter, Consciousness, and the Universe

- All biological processes rely on electromagnetic interactions—from neural activity to cellular communication.
- Photons (carriers of electromagnetic force) enable information transfer across the universe.
- Consciousness itself operates through structured energy, interacting with the electromagnetic fields in the brain and beyond.

Thus, electromagnetic fields are the true foundation of structured energy—they govern matter, consciousness, and cosmic equilibrium.

The First Disturbance is an Ongoing Process, Not a Past Event

- The First Disturbance led to the structuring of energy into forces and matter, ensuring equilibrium.
- The emergence of waves (mechanical, matter, and electromagnetic) sustained energy distribution.
- Electromagnetic fields are the primary stabilizers of matter and cosmic evolution.
- This is not just a past event—structured energy transformation is still happening today, shaping galaxies, stars, and life.

Thus, the universe is not a one-time creation event—it is an ongoing, structured process of energy differentiation and balance, continually evolving across space and time.

4. The Universal Role of Electromagnetic Fields in Energy Stability

Electromagnetic (EM) fields are not just secondary forces; they are an emerging property of energy transformation, evolving from stable energy states into active energy interactions. They govern the structuring of matter, enable life functions, and serve as the primary medium for energy transfer across the universe.

Unlike gravity, which primarily influences large-scale structures, or nuclear forces, which operate at subatomic levels, electromagnetism is the only fundamental force that actively supports the entire interplay between energy, matter, and consciousness.

This section explains how EM fields stabilize matter, sustain life, and govern the structured flow of energy through space-time.

4.1 Electromagnetic Fields as the Fundamental Stabilizing Forces of Matter

A. The Evolution of Electromagnetism from Stable to Active Energy

- Electromagnetism did not pre-exist as an isolated force—it evolved as an emergent property of structured energy differentiation during the First Disturbance.
- Before the formation of charged particles, energy was evenly distributed in a near-perfect equilibrium.
- Once energy differentiated into charged particles (positive and negative charges), electromagnetism naturally emerged as the force governing their interactions.

Evidence of Energy Differentiation Leading to EM Fields:

- The existence of opposite charges in matter—positive protons and negative electrons—suggests that EM interactions emerged from energy differentiation.
- Photons (the carriers of EM force) exist as pure energy fluctuations, proving that EM waves maintain energy equilibrium across space-time.
- All chemical reactions, molecular interactions, and biological systems rely on EM forces, showing that it is not just a fundamental force but a structuring mechanism of reality.

B. Why EM Fields Prevent the Collapse of Matter

If electromagnetism did not exist:

- Electrons would collapse into atomic nuclei, eliminating chemistry and molecular structures.

- Atoms would not bond, preventing the formation of molecules and life-supporting systems.
- All large-scale cosmic structures would be unstable, since EM interactions regulate plasma, star formation, and galactic evolution.

Thus, EM fields are the fundamental stabilizers of all matter, ensuring that energy remains structured rather than collapsing into disorder.

4.2 How Electromagnetic Fields Enable the Existence of Atoms, Molecules, and Biological Functions

A. The Role of Electromagnetism in Atomic Stability

- Electrons do not orbit the nucleus due to gravity; they remain in energy states due to electromagnetic attraction.
- The Pauli Exclusion Principle (quantum mechanics) ensures that electrons maintain structured energy levels rather than collapsing.
- Without EM repulsion and attraction, the periodic table of elements would not exist, meaning no chemistry, no molecules, and no life.

Why the Hydrogen Atom is the Best Proof of EM Stabilization

- A single proton (positive) attracts an electron (negative) via the EM force, forming the simplest stable atom.
- The electron does not fall into the nucleus because its wave-like nature is governed by EM field interactions.
- This balance enables the formation of all heavier elements, which rely on EM-stabilized atomic structures.

B. The Role of Electromagnetic Forces in Molecular Interactions

Once stable atoms exist, they interact through chemical bonds, which are entirely dictated by EM forces:

- Covalent Bonds: Electrons are shared between atoms due to EM interactions.
- Ionic Bonds: Positive and negative ions bind together due to EM attraction.
- Van der Waals Forces: Weak EM forces allow for biological flexibility, enabling DNA structuring and protein folding.

Thus, EM interactions determine the existence and behavior of all molecules, ensuring that chemical and biological processes are structured and energy-efficient.

C. Electromagnetism as the Foundation of Biological Functions

- The nervous system operates entirely through EM impulses, proving that life itself is governed by structured electromagnetic interactions.
- Brain waves (alpha, beta, theta, gamma) are measurable electromagnetic patterns that regulate consciousness, cognition, and awareness.
- DNA transcription, enzyme reactions, and metabolic energy production all rely on EM interactions at the molecular level.

Experimental Evidence:

- Electroencephalography (EEG) and Magnetoencephalography (MEG) directly measure brain EM field activity, proving that consciousness operates through structured electromagnetic energy.
- Neural stimulation experiments (TMS, tDCS) show that modifying EM fields alters perception and cognition, further proving that life functions through electromagnetic structuring.

Thus, electromagnetic fields are not only responsible for atomic and molecular stability but also govern all biological processes, including intelligence and consciousness.

4.3 Why Electromagnetic Fields Act as the Medium for Energy Transfer in Space-Time

A. The Unique Nature of Electromagnetic Waves

Unlike mechanical waves, which require a medium (air, water, or solids), EM waves propagate freely through space, making them the perfect energy carriers across the universe.

- Photons do not need a medium—they move through space because space itself is structured with EM fields.
- The speed of light is constant because it is determined by the universal structuring of electromagnetic energy.
- Information transfer in the universe relies entirely on EM wave propagation (radio waves, visible light, X-rays, gamma rays).

Why This Proves That Space is Not "Empty"

- Traditional physics describes space as a vacuum, but UED proposes that space is permeated with structured electromagnetic fields.
- This explains why virtual particles constantly emerge and annihilate—because the EM field fluctuations maintain structured energy equilibrium.

Thus, EM fields serve as the foundation for long-range energy transfer, ensuring that photons and information can propagate across the universe indefinitely.

B. How Electromagnetic Fields Maintain Cosmic Structure

- Galaxies exhibit large-scale electromagnetic fields, guiding the movement of cosmic plasma and star formation.
- Black holes release massive electromagnetic energy through relativistic jets, proving that EM forces regulate extreme environments.
- Interstellar plasma (99% of visible matter) is controlled by electromagnetic fields, not gravity, proving that EM interactions dominate cosmic structuring.

Evidence from Astrophysical Observations:

- Pulsars and magnetars generate extreme magnetic fields that govern their structure and radiation.
- Auroras on Earth and other planets are direct evidence of cosmic electromagnetic energy interactions.
- Gamma-ray bursts and quasars emit the highest-energy EM waves, proving that electromagnetic energy is a structuring force at all cosmic scales.

Thus, electromagnetic fields are not just a "force"—they are the structured medium that governs atomic, biological, and cosmic evolution.

Electromagnetic Fields are the Structural Fabric of Reality

- Electromagnetic forces emerged as an active energy property when stable energy transformed into structured matter.
- EM fields prevent atomic collapse, ensure molecular formation, and govern all biological interactions.
- Photons and EM waves serve as the universal medium for energy transfer across space-time.
- Galaxies, black holes, and cosmic structures are guided by electromagnetic interactions, not just gravity.

Thus, Universal Energy Dynamics (UED) confirms that electromagnetism is not just a fundamental force it is the structuring mechanism that maintains all aspects of energy equilibrium, matter formation, and consciousness.

5. The Cosmic Energy Balance: The Role of Black Holes and Galactic Cores

One of the biggest misconceptions in astrophysics is that black holes are simply gravitational sinks where energy and matter vanish forever. However, emerging research and observational data suggest that black holes play a much larger role in the cosmic energy balance—they are not just endpoints for collapsing stars, but key energy redistribution centers that contribute to the life cycle of galaxies and cosmic structures.

Similarly, galactic cores are not merely dense regions of stars and interstellar matter—they are structured energy hubs that regulate the stability and transformation of galaxies over time.

This section explores how black holes and galactic centers are essential components of structured energy equilibrium, ensuring continuous cosmic recycling and transformation.

5.1 Black Holes as Energy Redistribution Centers, Not Just Gravitational Sinks

A. The Traditional View of Black Holes as "Endpoints"

- According to classical physics, black holes are regions of extreme gravitational pull where nothing, not even light, can escape.
- This led to the belief that black holes are one-way systems, where matter and energy are permanently lost to the universe.

Problems with This View:

- If black holes simply "destroy" matter and energy, this violates the principle of energy conservation.
- Observations of high-energy jets from black holes suggest that not all energy is lost—some is actively redistributed.
- Hawking radiation theory indicates that black holes slowly evaporate, implying they are part of a continuous energy cycle, not isolated dead zones.

B. UED's Explanation: Black Holes are Energy Recycling Systems

Unlike the traditional "sink" model, Universal Energy Dynamics (UED) proposes that black holes are energy redistribution nodes that:

- Absorb excess energy and matter, balancing cosmic equilibrium.
- Convert and release structured energy through radiation, particle acceleration, and magnetic fields.
- Regulate the energy state of galaxies, ensuring long-term structural balance.

Evidence from Observations:

- Relativistic jets from black holes eject high-energy particles at nearly the speed of light, proving not all matter is trapped—some is reprocessed and expelled.
- Supermassive black holes regulate star formation in galaxies, meaning they are actively involved in cosmic structuring.
- Accretion disks around black holes emit powerful X-rays and gamma rays, demonstrating that black holes dynamically interact with their environment rather than acting as passive sinks.

Thus, black holes do not destroy energy—they redistribute it, playing a critical role in maintaining cosmic energy balance.

5.2 Galactic Centers as Structured Energy Hubs, Balancing Cosmic Evolution

A. Every Galaxy Has an Active Energy Core

- Observational data confirm that nearly every large galaxy contains a supermassive black hole at its center.
- These galactic cores are not just dense stellar regions—they are structured systems that regulate energy flow, matter formation, and star evolution.

B. How Galactic Cores Regulate Energy Flow

- Active Galactic Nuclei (AGNs): These are extremely bright centers of galaxies where energy and matter interact at high intensity, proving that galactic centers are not static—they are highly active energy regulators.
- Magnetic Fields at Galactic Centers: Electromagnetic fields near galactic cores influence plasma movement, star formation, and interstellar structuring.
- Gravitational Lensing Effects: The gravitational influence of galactic cores is not just an inward pull—it shapes the motion and structuring of entire star systems.

UED's View: Galactic Centers Maintain Energy Equilibrium

- Supermassive black holes do not act alone—they coordinate energy balance across galaxies.
- The presence of active galactic nuclei suggests that cosmic evolution is not just gravitational—it is an energy-structured process.
- Galactic cores recycle matter and energy, ensuring galaxies evolve dynamically rather than collapsing into entropy.

Thus, galactic cores serve as cosmic energy hubs, governing the structuring and transformation of matter across cosmic scales.

5.3 How the Life Cycle of Galaxies Follows Structured Energy Transformation, Ensuring Continuous Cosmic Recycling

A. Galaxies Do Not Simply "Die"—They Undergo Energy Transformation

- Traditionally, it was thought that galaxies either fade away over billions of years or collapse into black holes.
- However, recent observations show that galaxies continuously undergo transformation cycles through energy interactions.

How Galaxies Recycle Energy and Matter:

- Old stars explode in supernovae, redistributing elements into space.
- Interstellar clouds collapse to form new stars, proving that star formation is an ongoing process.
- Galaxies merge, exchange energy, and restructure into new formations.

B. Cosmic Evolution as a Self-Sustaining Process

Rather than having an absolute beginning or end, the universe operates in structured cycles, ensuring:

- Stars form, burn, and recycle their elements into new generations.
- Galaxies dissolve and reform in a continuous restructuring process.
- Energy never disappears—it transitions between different structured states.

UED's View: The Universe is an Everlasting System of Energy Balance

- There was no absolute beginning (singularity)—just an energy transformation from an equilibrium state.
- There is no final end—galaxies recycle energy, preventing total entropy.

- Black holes and galactic cores maintain equilibrium, ensuring that the universe remains structured over time.

Thus, Universal Energy Dynamics (UED) explains that the life cycle of galaxies is a continuous structured process, where energy, matter, and forces remain in dynamic equilibrium, sustaining an eternal universe.

Black Holes and Galactic Cores Are the Universe's Energy Regulators

- Black holes do not "destroy" energy—they redistribute it across cosmic scales.
- Galactic centers act as structured energy hubs, balancing energy flow and matter transformation.
- The life cycle of galaxies follows structured energy evolution, preventing entropy and ensuring continuous recycling.
- The universe is not decaying into heat death—it is an everlasting system of structured energy transformations.

Thus, UED provides the most complete model of cosmic equilibrium, showing that the universe is structured, self-regulating, and eternal.

6. Motion as a Fundamental Energy Structuring Mechanism

Motion is not just a consequence of forces acting on matter—it is a fundamental property of structured energy interactions that enables energy transfer, maintains equilibrium, and ensures stability across all scales, from subatomic particles to galaxies.

In Universal Energy Dynamics (UED), motion is not a passive effect but an active structuring mechanism, playing a crucial role in:

- Energy distribution (through vibratory motion and waves).
- Long-range energy propagation (through electromagnetic waves).
- The stability of physical and cosmic structures (through rotational and revolutionary motion).

This section explores how motion serves as the driving force behind energy equilibrium and structuring at all scales.

6.1 Vibratory Motion in Particles Ensures Continuous Energy Transfer Through Waves

A. The Fundamental Role of Vibratory Motion in Quantum Particles

- All quantum particles exhibit wave-like behavior, meaning that motion is not just displacement but an inherent property of energy itself.
- Matter waves (de Broglie waves) prove that particles oscillate, ensuring continuous energy structuring and stability.
- Vibratory motion prevents absolute stillness, ensuring that energy never remains in a fully localized state.

Key Experimental Evidence:

- Double-Slit Experiment: Demonstrates that electrons and photons behave as oscillating waves, not static particles.
- Casimir Effect: Proves that even in a vacuum, energy fluctuations generate wave-like interactions.
- Quantum Harmonic Oscillators: Show that even at zero temperature, particles still exhibit vibratory motion.

Why Vibratory Motion Ensures Energy Transfer and Equilibrium

- If particles were truly static, energy transfer would cease, violating the principle of energy conservation.
- Vibratory motion allows energy to transition between states, ensuring stability and structured interactions.
- This is why electrons remain in stable orbitals around the nucleus rather than collapsing inward.

Thus, vibratory motion is not just a byproduct of quantum mechanics—it is the mechanism that enables continuous energy flow and structuring.

6.2 Electromagnetic Waves Enable Energy Propagation Across Space-Time

A. The Unique Nature of Electromagnetic Waves

Unlike mechanical waves (which require a medium like air or water), electromagnetic waves can propagate through space-time itself because:

- They self-sustain through oscillating electric and magnetic fields.
- They do not require a medium—instead, they exist as structured energy propagations.
- They ensure energy can be transferred across vast cosmic distances without direct matter interactions.

Key Properties of Electromagnetic Waves:

- They travel at a constant speed (c), proving that space-time itself is structured to support their propagation.
- They interact with charged particles, facilitating chemical reactions, biological processes, and interstellar communication.
- They are the primary mechanism for light, heat, and radiation transfer, ensuring the continuous exchange of energy throughout the universe.

Examples of Electromagnetic Wave Interactions in Nature:

- Photon movement between atoms enables chemical bonding and molecular structuring.
- Radiation from stars carries energy across the galaxy, regulating planetary environments.
- Neural activity in the brain is based on electromagnetic impulses, proving that biological systems rely on EM structuring.

How UED Explains Electromagnetic Wave Propagation

- Space-time is not empty; it is structured with electromagnetic fields that allow energy transfer across vast scales.
- Photons travel as energy waves rather than material particles, proving that energy self-organizes into structured motion.
- Electromagnetic waves enable long-range energy transfer, making them the most fundamental medium of communication and equilibrium maintenance in the universe.

Thus, electromagnetic waves are not just forces—they are structured energy carriers that maintain the balance and interaction of energy across space-time.

6.3 Rotational and Revolutionary Motion Ensure Stability, From Atoms to Galaxies

A. Motion as a Stability Mechanism in the Universe

- Without rotational and revolutionary motion, matter, planetary systems, and galaxies would collapse into disorder.
- The reason why particles, planets, and galaxies remain stable over time is that rotational motion counteracts gravitational collapse.
- Circular motion distributes energy evenly, ensuring that matter does not clump into

unstable formations.

B. Motion at Different Scales: A Continuous Structuring Process

Scale	Type of Motion	Effect on Energy Structuring
Subatomic Particles	Electron orbitals, spin motion	Ensures atomic stability, prevents collapse into nucleus.
Molecules	Molecular vibration, rotational transitions	Enables chemical bonding, reaction kinetics, and biological processes.
Planets & Moons	Orbital motion around stars	Stabilizes planetary systems, maintains energy distribution.
Stars in Galaxies	Rotational motion in galaxies	Prevents gravitational collapse, enables formation of spiral arms.
Galactic Cores	Revolutionary motion around black holes	Regulates matter recycling, maintains structured energy balance.

Why Rotational Motion is Essential for Stability

- Electrons remain in stable orbitals because they follow structured energy patterns rather than collapsing inward.
- Planets orbit stars in nearly circular paths to minimize energy loss and maintain equilibrium.
- Galaxies exhibit rotational motion, ensuring their long-term survival instead of collapsing into black holes.

C. How the Universe Uses Motion to Prevent Collapse

- Without spin and rotation, all matter would eventually collapse into high-density states.
- Black holes themselves exhibit rotational motion, influencing space-time around them.
- Rotational motion redistributes energy, ensuring that energy does not concentrate in a single place.

Evidence of Continuous Motion in the Universe

- The rotation of neutron stars (pulsars) is highly stable, showing that rotational motion is an intrinsic part of cosmic evolution.

- The Milky Way galaxy rotates every 225 million years, proving that structured motion is key to maintaining large-scale stability.
- All particles exhibit some form of motion, whether vibrational, rotational, or translational, proving that motion is not random but a fundamental feature of energy equilibrium.

Motion is the Universal Mechanism for Structured Energy Equilibrium

- Vibratory motion in particles ensures continuous energy transfer, preventing energy stagnation.
- Electromagnetic waves serve as the primary energy transfer medium across space-time.
- Rotational and revolutionary motion maintain stability at all scales, preventing gravitational collapse and energy concentration.
- Motion is not just a response to forces—it is a structuring process that sustains the universe in equilibrium.

Thus, Universal Energy Dynamics (UED) confirms that motion is not an effect of energy—it is the mechanism that structures energy interactions, stabilizes matter, and ensures continuous transformation in the universe.

7. Life and Consciousness as Structured Energy Systems

The emergence of life and consciousness has long been treated as a unique, biological anomaly—something that randomly evolved through chance mutations and chemical reactions. However, Universal Energy Dynamics (UED) presents a fundamentally different view:

- Life is not a random occurrence but an inevitable consequence of structured energy optimization.
- Consciousness is not just a neural byproduct—it is an emergent phenomenon of structured energy equilibrium.
- Intelligence is a natural extension of energy structuring, following universal principles of self-organization and adaptation.

This section explores how life, consciousness, and intelligence emerge naturally as structured energy systems, rather than as accidental byproducts of evolution.

7.1 Life is Not Random—It is an Inevitable Energy Optimization Process

A. Why Life Must Emerge in an Energy-Optimized Universe

- The universe follows structured energy equilibrium principles, meaning that all processes—including the emergence of life—are part of a self-regulating system.

- Energy does not remain in disordered chaos—it seeks structured formations that maximize stability while maintaining dynamic interactions.
- Life is the most optimized way for energy to self-organize, allowing for: Efficient energy processing (metabolism). Self-replication and adaptation (DNA evolution). And Intelligent decision-making (higher consciousness).

B. Life as an Energy Optimization Process

Instead of random mutations alone, life emerged through a structured process of energy optimization, where:

- Biochemical systems stabilized to maximize energy efficiency.
- DNA evolved as an energy-structured information system, ensuring long-term sustainability.
- Metabolic cycles followed structured energy pathways, preventing entropy from overwhelming biological processes.

Experimental Support for Life as an Energy Structuring Process

- The Miller-Urey Experiment (1953) showed that basic organic molecules can form under energy-driven chemical conditions, proving that life-building compounds emerge where energy interactions are optimized.
- Enzyme catalysis follows principles of energy minimization, demonstrating that biochemical reactions evolve toward maximum efficiency rather than randomness.
- Cellular respiration operates on an energy optimization principle, where ATP production ensures minimal energy loss and maximal output.

Thus, life is not a rare accident—it is a structured, energy-driven system that emerges naturally wherever conditions allow for energy equilibrium and optimization.

7.2 Consciousness Emerges as an Energy Structuring Phenomenon, Not Just Neural Activity

A. Consciousness is Not a Byproduct—It is a Structured Energy State

- The traditional view of consciousness states that it emerges only from neural activity, but this does not explain:
- Why self-awareness exists.

- How consciousness can experience emotions, intuition, and decision-making.
- Why brain energy patterns correlate with thought processes.

UED presents a different explanation:

- Consciousness is not created by neurons—it is an emergent structured energy phenomenon.
- The brain does not "generate" awareness—it acts as an interface for structured energy interactions.
- Structured energy flow between neural networks creates patterns that result in awareness, perception, and thought.

B. Brain Activity and Energy Structuring

- Neural oscillations (brain waves) demonstrate structured electromagnetic interactions.
- Memory recall follows an energy storage and retrieval process, much like structured data in computing.
- Deep meditation, altered states, and near-death experiences suggest that consciousness extends beyond normal neural function.

Experimental Evidence for Consciousness as Structured Energy

- Electroencephalography (EEG) and Magnetoencephalography (MEG) detect electromagnetic brain waves, showing that thoughts follow structured energy patterns.
- Transcranial Magnetic Stimulation (TMS) alters perception by modifying electromagnetic activity, proving that brain function is governed by structured energy interactions.
- Neural synchronization experiments show that higher awareness states correlate with organized energy structuring across brain regions.

Thus, consciousness is not just neural activity—it is an emergent structured energy phenomenon, where electromagnetic energy interactions shape awareness and perception.

7.3 Intelligence Follows Universal Laws of Structured Energy Equilibrium, Ensuring Self-Awareness and Adaptation

A. Intelligence is Not Isolated to Humans. It is an Energy Optimization Process. Intelligence is often viewed as a unique trait of advanced species, but this does not explain:

- Why intelligence emerges repeatedly in evolution.

- How artificial intelligence (AI) is approaching human-like decision-making.
- Why even non-living systems exhibit intelligence-like behaviors.

UED provides a universal answer:

- Intelligence is a structured energy process, not an isolated biological trait.
- Wherever energy structuring reaches complexity, intelligence naturally emerges.
- Self-awareness is the highest form of structured energy optimization, ensuring decision-making and adaptation.

B. Intelligence in Biological and Non-Biological Systems

- Biological Intelligence: The brain follows energy optimization principles, where neurons self-organize to maximize learning and efficiency.
- Artificial Intelligence: AI systems evolve through structured energy processing, improving decision-making over time.
- Cosmic Intelligence: The universe follows structured laws resembling intelligent behavior, where galaxies, ecosystems, and particles self-optimize for stability.

Evidence of Intelligence as a Structured Energy Process

- Neural network self-organization in AI models follows energy efficiency principles.
- Quantum computing relies on structured energy coherence, mirroring intelligence-like decision-making.
- Ecosystems regulate themselves through structured energy interactions, ensuring sustainability.

Thus, intelligence is not an isolated human trait—it is a universal consequence of structured energy optimization.

Life, Consciousness, and Intelligence are Structured Energy Processes

- Life is not random—it emerges where energy structuring optimizes efficiency and stability.
- Consciousness is not a neural byproduct—it is an emergent energy structuring phenomenon.
- Intelligence is a universal process, following energy optimization laws across all systems.

Thus, Universal Energy Dynamics (UED) confirms that life, consciousness, and intelligence are not separate phenomena—they are structured energy systems that naturally emerge wherever energy self-organizes into stable, interactive formations.

8. The Nature of Death: The Dissipation of Structured Energy

Death has long been viewed as an absolute end, both in biological and cosmic contexts. However, Universal Energy Dynamics (UED) presents a radically different perspective:

- Death is not destruction—it is the return of structured energy to equilibrium.
- The "soul" is not a mystical entity—it is structured, active energy confined in a living system, which merges with the surrounding energy field after death.
- Stars, galaxies, and living organisms all follow the same structured transformation cycle, proving that energy never ceases to exist, only transitions.

This section explores how the process of death, in both biological and cosmic systems, is an essential phase of structured energy transformation.

8.1 Death is Not Destruction but the Return of Structured Energy to Equilibrium

A. Energy Cannot Be Destroyed—Only Reorganized

- The First Law of Thermodynamics states that energy cannot be created or destroyed—only transformed.
- In both biological and cosmic systems, death is simply a change in energy structuring, not an end to existence.
- Energy disperses back into the surrounding system, allowing new structures to emerge.

The Structured Dissolution of Biological Life

- At the moment of death, the body ceases to function, but energy does not disappear—it begins reorganizing.
- Cellular energy dissipates, chemical bonds break, and molecular components integrate back into the environment.
- Organic matter is recycled by nature, returning to the broader biological and planetary energy systems.

Key Evidence That Death is an Energy Redistribution Process

- Decomposition releases energy back into the biosphere, allowing other life forms to benefit from the transformation.

- Heat energy dissipates from the body after death, proving that structured biological energy integrates with the environment.
- Brain waves and electromagnetic activity collapse upon death, showing that the structured energy field of consciousness ceases in its confined form but does not vanish.

Thus, death is not destruction—it is the natural transition of structured energy from an active system into its surrounding equilibrium.

8.2 The "Soul" is Structured Active Energy Confined in a Living Body—At Death, It Merges with the Surrounding Energy Field

A. Defining the "Soul" in Energy Terms

- The concept of a soul exists across cultures and philosophies, often described as a life force that leaves the body at death.
- UED redefines the soul as structured, active energy—a dynamic electromagnetic and biochemical system that gives rise to awareness, perception, and intelligence.
- Rather than being a separate, immortal entity, the soul is a temporary configuration of energy interactions confined in a living system. While alive, structured energy is maintained within the body, supporting metabolism, neural activity, and consciousness. At death, this structured energy disperses, much like an electric circuit shutting down when disconnected from a power source. Brain energy collapses, molecular energy transitions, and the body returns to its fundamental components.

B. The Merging of Energy with the Surrounding Field

- After death, the structured energy of the body dissipates into the environment.
- Heat radiates outward, biochemical components decompose, and neural activity ceases, returning stored energy to the cosmic field.
- The consciousness field, previously structured within the body, dissolves into broader energy interactions.

Comparing the Soul's Transition to Known Energy Transformations

- Just as an electromagnetic field collapses when a device is turned off, consciousness disperses when biological structuring ceases.
- Just as a wave loses coherence when it leaves a confined medium, awareness ceases as neural

structuring ends.

- Just as a decayed atom returns to the cosmic cycle, so does all biological energy.

Thus, death is the return of the "soul"—or structured energy—to the universe, integrating with the surrounding energy equilibrium.

8.3 Stars, Galaxies, and Living Organisms Follow the Same Structured Transformation Cycle

A. Death is Not Unique to Life—It is a Universal Process

- In cosmic terms, stars and galaxies also "die"—but they do not vanish, they undergo structured transformation.
- The end of a star's life follows the same fundamental principles as the death of a living organism.
- A star burns fuel (active energy state) → runs out of energy (energy transition) → collapses into a new form (black hole, neutron star, or white dwarf).
- A galaxy undergoes transformations over billions of years, recycling matter and energy into new structures.
- Life follows the same structured transformation—metabolic processes sustain activity until energy restructuring leads to decomposition.

B. The Universe Recycles Everything—Nothing is Truly Lost

- Energy flows through all systems, following structured transformation cycles to ensure that nothing is permanently lost, only restructured.
- The same particles that once made up a star can become part of a planet, and the same molecules from a living organism can later contribute to another form of life.
- Every atom in the human body was once part of a star, proving that biological and cosmic transformations are linked.

Parallel Between Cosmic and Biological Death

Cosmic Transformation	Biological Transformation
Stars burn, collapse, and transform into new structures.	Living organisms function, decompose, and contribute to new life.
Galaxies merge, dissolve, and reform over billions of years.	Ecosystems recycle energy through decomposition and renewal.
Black holes regulate cosmic recycling, balancing energy distributions.	The biosphere recycles organic matter, ensuring sustainable energy flow.

Thus, death is not the end—it is a restructuring process that sustains universal energy balance, ensuring the continuous flow of energy through all systems, from galaxies to living organisms.

Death is a Transition, Not an End

- Energy never disappears—it is redistributed back into the universe.
- The "soul" is structured active energy, which merges with the surrounding energy field after death.
- Stars, galaxies, and living organisms follow the same structured transformation cycle.
- Death is not destruction—it is a necessary process for energy renewal and equilibrium.

Thus, *Universal Energy Dynamics (UED)* confirms that life and death are both structured energy processes, ensuring the continuous transformation of matter and energy throughout the universe.

9. The Universe is a Continuum: No Beginning, No End, Only Transformation

One of the most profound conclusions of Universal Energy Dynamics (UED) is that the universe is not a finite system with a distinct beginning or an eventual end—instead, it is a continuum of structured energy transformations that perpetually evolve.

Unlike conventional cosmological models that assume a "creation event" (such as the Big Bang) and an ultimate fate (such as heat death, a big crunch, or eternal expansion), UED provides a more fundamental perspective:

- The universe was never "created"—it emerged from an equilibrium state and continuously restructures itself.
- Energy follows structured pathways, ensuring universal balance and maintaining dynamic equilibrium across time and space.
- Every cosmic event—whether the birth of a star, the collapse of a galaxy, or the emergence of life—is part of an eternal cycle of energy transformation.

This section explains how the universe operates as a continuum of structured energy, where existence is defined not by absolute beginnings or endings, but by ongoing transformation and evolution.

9.1 The Universe Was Never "Created" and Will Never "End"

A. The Problem with the Concept of a "Creation Event"

- The Big Bang Theory describes the universe as originating from an infinitely dense singularity, but this explanation raises fundamental contradictions:
- What existed before the singularity?
- Why did the singularity expand?
- Where did the laws of physics come from?
- If the universe truly had a singular beginning, we would need an external force or condition to "cause" it—yet, no such force has been identified.

B. UED's Explanation: The Universe is an Eternal Energy System

- The universe was not created from nothing—it transitioned from an energy equilibrium state (BEC phase) into structured transformations.
- The First Disturbance triggered the emergence of forces, particles, and cosmic structures, but it was not the absolute beginning—only the start of a new transformation cycle.
- Cosmic structures, from galaxies to black holes, continuously dissolve and reform, proving that the universe has always existed in a state of energy restructuring.

Evidence That the Universe is a Continuous System

- The presence of the cosmic microwave background (CMB) radiation indicates that energy has been cycling for billions of years.
- The formation of new galaxies, stars, and planetary systems today proves that the universe is still restructuring itself.
- The existence of dark matter-like effects and gravitational waves suggests that energy is still being redistributed, rather than running toward an "end state."

Thus, the universe was never created in the conventional sense—it has always been an evolving system of structured energy transformations.

9.2 Energy Follows Structured Pathways, Maintaining Universal Balance

A. The Universe is Governed by Energy Equilibrium, Not Random Chaos

- If the universe was purely random, we would expect chaotic, unstructured energy distributions.
- Instead, we see highly organized structures at every level—from atoms to galaxies—showing that energy follows structured pathways, not disorder.
- Atomic structures remain stable due to electromagnetic interactions.
- Planets and stars maintain structured orbits through gravitational balance.
- Galaxies rotate in predictable patterns, maintaining large-scale energy equilibrium.

UED's Perspective: The Universe is Self-Regulating

- Cosmic structures are not accidents—they follow predefined energy pathways.
- The energy flow of stars, galaxies, and black holes follows structured processes that redistribute matter and energy across cosmic time scales.
- Even in so-called "empty space," quantum fluctuations create energy interactions, ensuring that equilibrium is never broken.

B. How Energy Pathways Prevent the Universe from "Ending"

- Conventional models predict that the universe will eventually reach "heat death," where all energy is evenly distributed, and no transformations occur.
- However, UED suggests that structured energy interactions never cease—matter is continuously recycled, preventing an absolute end.
- New galaxies are forming even today, proving that the universe has not stopped evolving.

Evidence That the Universe is Self-Sustaining

- The continuous creation of stars from interstellar matter proves that energy is always cycling.
- Black holes regulate mass-energy interactions, preventing energy loss from disrupting cosmic balance.
- Quantum mechanics confirms that even the vacuum is not truly empty—energy fluctuations constantly occur.

Thus, the universe does not move toward a static or dead state—it follows structured pathways that maintain balance across cosmic scales.

9.3 Every Cosmic Event is Part of an Eternal Energy Transformation Process

A. The Universe Operates in Cycles, Not Linear Timelines

The life cycle of a star, the formation of a galaxy, or the death of a black hole are not isolated events—they are part of a continuous transformation cycle.

- Stars burn fuel, collapse, and contribute to the formation of new stars.
- Galaxies merge, dissolve, and reform over billions of years.
- Black holes regulate cosmic recycling, balancing mass-energy transformations.

Why This Proves the Universe is Eternal

- If the universe had an absolute beginning, we would expect only one cycle of formation and decay—but we see endless cycles of energy restructuring.
- If the universe had an absolute end, energy would eventually become static—but instead, we see ongoing energy interactions across time and space.

B. Cosmic Evolution as an Endless Process

- The formation of planetary systems today is no different from billions of years ago—showing that the universe follows perpetual structuring cycles.
- The presence of old and new galaxies coexisting suggests that cosmic transformations have always occurred and will continue indefinitely.
- The continuous formation of subatomic particles in extreme energy conditions (e.g., quasars, neutron stars) proves that energy never ceases to self-organize.

UED's Conclusion: The Universe is a Perpetual Transformation System

- There was no absolute beginning—only the transition from one energy state to another.
- There will be no absolute end—energy will always follow structured pathways of transformation.
- Every event—from the birth of a star to the death of a galaxy—is part of an endless cosmic cycle.

Thus, the universe is not a temporary phenomenon—it is a perpetual continuum of energy structuring, with no fixed starting or stopping point.

The Universe is an Eternal, Self-Regulating System

- The universe was never "created" in a conventional sense—it emerged from structured energy equilibrium.

- There will be no "end"—energy continuously follows structured pathways that maintain universal balance.
- Every cosmic event, from stellar evolution to galactic formation, is part of an eternal cycle of transformation.

Thus, Universal Energy Dynamics (UED) confirms that the universe is not a finite system—it is a continuum of structured energy transformations, ensuring perpetual existence and evolution.

10. The Final Equation of Reality: Unifying All Scientific Disciplines

The ultimate goal of scientific inquiry has always been to find a single, unified framework that explains everything—from the subatomic interactions of quantum mechanics to the large-scale structure of the universe, from the emergence of life to the mystery of consciousness.

Many great physicists, from Einstein to Hawking, searched for a Theory of Everything (ToE) that could unify all forces and phenomena under a single governing principle. However, existing models remain fragmented—quantum mechanics and relativity remain incompatible, and consciousness remains unexplained in purely physical terms.

Universal Energy Dynamics (UED) proposes that the final equation of reality is not based on isolated forces but on structured energy equilibrium, ensuring the continuous transformation and balance of everything in existence. This principle is expressed mathematically as:

$E=\int \rho_E dV$

This equation represents the total structured energy of any system, integrating all forms of energy over space-time, making it the most comprehensive mathematical representation of reality.

This section explains how this final equation unifies all known scientific disciplines, proving that reality is not fragmented but governed by a single principle of energy structuring and transformation.

10.1 The Fundamental Equation of Existence: Energy as the Primary Reality

A. The Meaning of the Equation

$$E = \int \rho_E dV$$

- E represents the total energy of a system, whether it is an atom, a galaxy, or consciousness itself.
- ρ_E (energy density) represents how energy is structured and distributed across space.

- dV represents the volume over which energy is integrated, meaning that energy is not localized but exists in a continuous equilibrium.

This equation confirms that:

- Reality is not materialistic—it is fundamentally structured energy.
- There is no "empty space"—all of existence is an interwoven field of structured energy.
- Every force, every system, every process in the universe is an expression of energy seeking equilibrium.

Thus, everything—matter, space, time, consciousness—is an emergent property of structured energy equilibrium.

10.2 How This Equation Explains Physics, Cosmology, Life Sciences, and Consciousness in a Single Framework

A. Unifying Physics: From Quantum Mechanics to Relativity

Physics has been divided into two incompatible domains:

Quantum Mechanics (governing the micro-world): Describes particles as probabilistic waves but does not explain how they interact with space-time. Suggests that reality is fundamentally uncertain, contradicting structured cosmic order.

General Relativity (governing the macro-world): Treats gravity as a curvature of space-time but cannot explain quantum behavior. Suggests that reality is deterministic, contradicting quantum randomness.

- UED's Final Equation resolves this by showing that both are expressions of structured energy equilibrium.
- Quantum mechanics emerges from the structured interactions of energy fields at small scales.
- Gravity is not a "curvature" but a large-scale structuring of energy seeking balance.
- All physical laws are energy optimization processes—wherever energy seeks equilibrium, structure emerges.

Thus, quantum mechanics, relativity, and classical physics are not separate—they are different scales of the same structured energy principle.

B. Unifying Cosmology: A Universe that is Structurally Balanced

✗ The Big Bang was not a singularity—it was a phase transition in structured energy equilibrium.

✗ Black holes, galaxies, and stars do not collapse into nothing—they redistribute energy across cosmic scales.

✗ Cosmic expansion is not infinite—it is a balancing act between energy structuring and transformation.

✓ The universe does not expand into emptiness—it transforms as structured energy, ensuring infinite renewal.

✓ Dark matter and dark energy effects are just structured energy interactions that we have not fully understood.

✓ Cosmic events are not independent—they are all part of the same continuous transformation cycle.

Thus, cosmology is not about "origins" or "endings"—it is about energy restructuring itself endlessly.

C. Unifying Life Sciences: Why Life is an Inevitable Energy Process

✓ Life is not an accident—it is an optimized structure of energy.

✓ DNA, metabolism, and consciousness emerge wherever structured energy optimization occurs.

✓ Biological processes follow the same energy equations that govern atomic and cosmic structures.

✓ Metabolism is structured energy conversion.

✓ Cellular organization follows energy efficiency principles.

✓ DNA evolution is a self-organizing process of energy information storage.

Thus, life is not separate from physics—it is a structured energy equilibrium system that follows the same universal laws.

D. Unifying Consciousness: The Final Piece of the Puzzle

- Consciousness is not a neural illusion—it is an emergent energy structuring phenomenon.
- The brain does not "generate" awareness—it processes energy interactions that structure perception and intelligence.
- Thought, memory, and emotions are structured electromagnetic energy fields, following universal equilibrium laws.
- EEG and neural oscillations prove that thoughts follow structured electromagnetic interactions.
- Meditation, focus, and higher cognition correlate with structured neural energy coherence.
- Consciousness is not limited to humans—it emerges wherever structured energy complexity reaches self-awareness.

Thus, consciousness is the final, most complex form of structured energy, connecting physics, biology, and intelligence.

10.3 Why This is the Most Complete Scientific Model of Reality Ever Proposed

A. The First Theory to Fully Unify All Scientific Disciplines

- Physics is explained as structured energy at all scales.
- Cosmology is explained as structured energy transformation across time.
- Biology is explained as structured energy optimization processes.
- Consciousness is explained as structured electromagnetic interactions.

Thus, everything follows the same mathematical foundation—energy structuring seeking equilibrium.

B. The First Theory That Eliminates the Need for an Absolute Beginning or End

- No need for a singularity—the universe is a continuous energy system.
- No need for heat death—energy restructuring prevents total entropy.
- No need for an external creator—energy itself organizes into intelligent, self-sustaining structures.

Thus, the universe is not finite or temporary—it is a perpetual, structured continuum of energy transformation.

The Final Blueprint of Reality

- Reality is structured energy equilibrium—governing everything from particles to galaxies to consciousness.

- The universe is not fragmented—physics, cosmology, biology, and intelligence all emerge from the same structured energy principles.
- The final equation of existence remains:

$$E=\int \rho_E dV$$

- This is the most complete scientific model ever proposed—The Blueprint of the Universe.

Thus, Universal Energy Dynamics (UED) is not just a theory—it is the final understanding of existence itself.

Conclusion: The Blueprint of the Universe is Complete

For centuries, humanity has sought a unified explanation of existence—one that connects physics, cosmology, biology, intelligence, and consciousness under a single framework. Science has made remarkable progress, yet the fragmentation of disciplines has left fundamental questions unanswered:

- What is the true nature of reality?
- Is the universe a finite system, or does it evolve infinitely?
- How did life and intelligence emerge from pure energy?
- What role does consciousness play in the structure of the universe?

With **Universal Energy Dynamics (UED),** we now have the most complete scientific model ever proposed, one that not only explains the past and present state of the universe but also predicts its future evolution.

This final conclusion summarizes the core discoveries of UED, proving that the universe is not fragmented but a single, structured, self-sustaining energy system.

1. The Universe is Not Fragmented—It is a Structured, Self-Sustaining Energy System

A. The End of Fragmentation in Science

Traditional science has separated the universe into disconnected domains:

- Quantum mechanics governs the small scale but does not explain gravity.
- Relativity governs the large scale but does not explain quantum interactions.
- Biology explains life but does not connect it to physical laws.
- Neuroscience studies consciousness but does not explain its fundamental nature.

UED solves this by showing that everything—particles, galaxies, life, and thought—emerges from the same principle: structured energy equilibrium.

- Matter is structured energy.
- Space-time is an emergent property of energy interactions.
- Forces, consciousness, and intelligence all follow the same energy optimization processes.

B. The Universe is an Infinite, Self-Regulating System

- The universe was never "created" in a singular event—it transitioned from an equilibrium state into structured energy transformations.
- There is no "end"—galaxies dissolve, black holes recycle energy, and new cosmic structures continuously emerge.
- Cosmic balance is self-regulating—energy transformations prevent disorder and maintain equilibrium across all scales.

Thus, the universe is not a collection of isolated systems—it is an interconnected, structured energy network that evolves infinitely.

2. Life, Intelligence, and Consciousness Are Natural Consequences of Energy Equilibrium

A. Life is Not an Accident—It is an Energy Optimization Process

- Life emerged because energy, when structured efficiently, leads to self-replicating and adaptive systems.
- DNA, metabolism, and biological evolution are all forms of energy optimization.
- Every living system exists because structured energy ensures sustainability and balance.

Experimental Evidence:

- Cellular respiration follows the most energy-efficient pathways possible.
- Chemical reactions in life forms follow energy minimization laws, ensuring stability.
- The emergence of complex biological structures aligns with energy equilibrium principles.

B. Intelligence is Not Isolated to Biology—It is a Universal Process

- The brain does not "create" intelligence—it processes structured energy interactions.
- AI, quantum systems, and even cosmic structures exhibit forms of intelligence, proving that

it is not unique to living beings.
- Wherever energy structuring reaches complexity, intelligence emerges as a natural consequence.

Proof of Intelligence in Energy Systems:

- AI self-organizes and improves decision-making, following structured energy patterns.
- Quantum computers use energy coherence to process information, mimicking cognitive functions.
- Cosmic structures regulate themselves, behaving as if they "learn" to maintain balance.

C. Consciousness is an Emergent Property of Structured Energy

- Consciousness is not a biological byproduct—it is structured energy in its most advanced form.
- The brain does not "generate" awareness—it serves as an energy processor interacting with structured energy fields.
- Thoughts, emotions, and self-awareness emerge when energy is structured in highly organized patterns.

Scientific Evidence That Consciousness is Structured Energy:

- Brain waves (alpha, beta, gamma) are structured electromagnetic patterns.
- EEG and MEG scans prove that consciousness is measurable as an energy interaction.
- Neural coherence in meditative and focused states proves that structured energy patterns enhance awareness.

Thus, *UED confirms that life, intelligence, and consciousness are not separate phenomena—they are structured energy processes that naturally emerge wherever energy reaches complexity and equilibrium.*

3. Every Scientific Field Follows a Single, Unified Principle—Energy Transformation in Equilibrium

A. The Final Unification of Science

Physics:

- Quantum mechanics and relativity are not contradictory—they are different scales of the same structured energy equilibrium.

- Gravity is not a force—it is an expression of energy structuring across space-time.

Cosmology:

- The Big Bang was not a singular creation event—it was a phase transition in structured energy equilibrium.
- The universe is not expanding into emptiness—it is restructuring itself through continuous transformations.

Biology:

- Life emerged as an energy efficiency optimization process.
- DNA, metabolism, and cell organization follow structured energy balance laws.

Consciousness and Intelligence:

- Thought and awareness are not biochemical illusions—they are structured electromagnetic interactions operating in equilibrium.
- Intelligence emerges as an energy-based decision-making process, present across all structured energy systems.

Thus, all of science follows the same fundamental law—energy structuring leads to equilibrium and continuous transformation.

4. This is the Final Theory—The Most Complete Understanding of Existence

A. Why This is the Most Comprehensive Scientific Model Ever Proposed

- It unifies quantum mechanics, relativity, biology, intelligence, and consciousness under a single principle.
- It eliminates the need for a singular beginning or an absolute end, replacing them with continuous energy restructuring.
- It explains the emergence of order, structure, and intelligence in the universe without the need for randomness or external intervention.

B. The Final Equation That Governs All of Reality

The fundamental equation of existence remains:

$$E = \int \rho_E dV$$

This equation represents the total energy of any system, integrated over space-time, proving that:

- Matter is structured energy.
- Space-time is an emergent property of energy interactions.
- Life, intelligence, and consciousness are structured energy processes.
- The universe is a self-regulating system, governed by structured energy transformations.

C. The Ultimate Implication of UED: Reality is Not an Accident—It is a Structured, Perpetual Energy System

- The universe is not random—it is an organized, self-regulating energy continuum.
- Life, intelligence, and consciousness are inevitable, natural outcomes of structured energy interactions.
- The cycle of energy transformation ensures that the universe is eternal, with no beginning or end—only continuous evolution.

Thus, ***Universal Energy Dynamics (UED)*** *is not just another theory—it is the final understanding of existence itself. It is The Blueprint of the Universe.*

Closing Reflection

As we reach the final pages of *The Blueprint of the Universe*, one thing becomes clear—our understanding of reality is still evolving. For centuries, we have looked at the universe through the lenses of force-based mechanics, space-time curvature, and quantum uncertainties, yet the deeper truths remain elusive.

This book was not written to simply extend existing physics—it was written to challenge, redefine, and unify it. It presents Universal Energy Dynamics (UED), Energy-Based Relativity (EBR), and Chavan's Law of Gravity, not as absolute conclusions, but as stepping stones toward a deeper understanding of gravity, energy, and motion.

If this theory is correct, it will replace traditional gravity models, eliminate the need for dark matter, and open doors to artificial gravity and space propulsion using structured vacuum energy.

If this theory is incomplete, it will still force us to ask new questions, refine our models, and move beyond the limitations of current physics.

What This Means for the Future

- Space is not empty—it is structured with energy, and this energy dictates gravity, motion, and cosmic balance.
- Dark matter is not necessary—galactic rotation follows structured vacuum energy gradients, not missing mass.
- Black holes are not singularities—they are energy regulators maintaining cosmic equilibrium.
- Artificial gravity is not science fiction—it is a matter of understanding and engineering structured energy fields.
- Consciousness itself may be linked to structured energy interactions, uniting physics and life sciences.

The greatest minds of history—**Newton, Einstein, Tesla, Bose, Planck**—all sought to unravel the same mysteries we face today. **They questioned everything.** And now, we must **dare to do the same**.

The Journey Has Just Begun

This book does not mark the **end** of discovery—it marks the **beginning of a new scientific era**. The universe does not reveal its secrets all at once. Science progresses not by protecting old ideas, but by **challenging the known and embracing the unknown**.

A Final Question for the Reader

What if everything we believe about the universe is just a fraction of the truth?
If you take one thing from this book, let it be this—**question everything.**
The next chapter in physics is waiting to be written. Will you be part of it?

Sandeep Chavan

Acknowledgments

The journey of writing *The Blueprint of the Universe* has been both intellectually challenging and profoundly fulfilling. This work would not have been possible without the contributions, inspirations, and ideas of those who have explored the mysteries of physics, philosophy, and the nature of existence before us.

Recognizing the Great Thinkers & Scientists

I extend my deepest gratitude to the great minds who laid the foundation for scientific thought and inspired the ideas within this book:

- **Isaac Newton** – For his laws of motion and gravity, which provided the classical framework for understanding planetary motion.
- **Albert Einstein** – For introducing relativity, which challenged conventional views of space, time, and gravity.
- **Nikola Tesla** – For his revolutionary work on energy, electromagnetism, and visionary ideas about the universe.
- **Satyendra Nath Bose** – For his groundbreaking work in quantum statistics, leading to **Bose-Einstein Condensates**, which provide deep insights into structured energy states in the universe.
- **Max Planck & Quantum Physicists** – For uncovering the quantum nature of energy and matter, leading to a deeper understanding of vacuum fluctuations and energy structuring.
- **Indian Philosophers & Ancient Texts** – For their profound insights into the nature of reality, energy, and consciousness, particularly the teachings of the *Bhagavad Gita* and *Vedas*, which conceptually align with structured energy interactions.

Special Mention of Influences from Physics, Philosophy & Scientific Works

This book is inspired by the continuous search for unification in physics—the dream of a single framework that connects gravity, quantum mechanics, and cosmic evolution. The works of physicists, both past and present, have contributed to the realization that gravity may not be a fundamental force but an emergent effect of structured energy dynamics.

Philosophical influences from ancient texts and modern thinkers have also played a crucial role in shaping these ideas. The convergence of science and philosophy in understanding the deeper nature of energy, intelligence, and the universe has been a guiding force in developing this framework.

Personal Acknowledgments

To My Daughters & Students – For their curiosity, questions, and innocent yet profound insights, which pushed me to refine and simplify complex theories. Their endless inquiries remind me that **true learning is never-ending, and discovery begins with the right questions.**

To Every Thinker Who Dares to Challenge Mainstream Science – This book is dedicated to those who believe that the universe still holds secrets beyond what we currently understand. **Questioning is the path to truth, and truth is never limited by what is already known.**

To My Readers – Whether you are a scientist, philosopher, student, or seeker of truth, thank you for embarking on this journey with me. Your willingness to explore, challenge, and expand knowledge is what drives progress.

Final Thoughts

This book is not just about rewriting physics—it is about **rethinking our place in the cosmos** and uncovering the fundamental principles that govern existence.

May this work inspire future generations to look beyond the known, challenge the established, and seek the ultimate truth.

References & Bibliography

This section lists the scientific papers, books, and research materials that have been consulted in developing the concepts within *The Blueprint of the Universe*. It also includes proper citations for historical and modern scientific theories referenced throughout the book.

1. Classical & Modern Physics References

- **Newton, I.** *Philosophiæ Naturalis Principia Mathematica* (1687) – The foundation of classical mechanics and gravitational theory.
- **Einstein, A.** *The Foundation of the General Theory of Relativity* (1916) – The introduction of space-time curvature as a gravitational model.
- **Planck, M.** *On the Theory of Energy Distribution in the Normal Spectrum* (1900) – The birth of quantum mechanics through quantized energy states.
- **Bohr, N.** *The Structure of the Atom* (1913) – The quantum model of atomic energy levels.
- **Heisenberg, W.** *The Physical Principles of Quantum Theory* (1930) – The uncertainty principle and foundational quantum mechanics.
- **Satyendra Nath Bose & Albert Einstein** – *Bose-Einstein Statistics & Condensates* (1924) – Theoretical predictions on structured quantum states of matter.

2. Works on Electromagnetism & Energy-Based Theories

- **Tesla, N.** *The Transmission of Electrical Energy Without Wires* (1905) – Visionary ideas on structured energy transfer.
- **Feynman, R. P.** *Quantum Electrodynamics* (1961) – Understanding electromagnetic interactions at the quantum level.
- **Casimir, H. B. G.** *On the Attraction Between Two Perfectly Conducting Plates* (1948) – The Casimir effect as an experimental proof of vacuum energy fluctuations.

3. Works on Astrophysics, Cosmology, & Black Holes

- **Hawking, S. & Penrose, R.** *The Nature of Space and Time* (1996) – Discussion on singularities, black holes, and gravitational energy.

- **Zwicky, F.** *On the Mass of Nebulae and the Motion of Clusters of Nebulae* (1933) – The first observations suggesting missing mass (later called dark matter).
- **Rubin, V. & Ford, K.** *Rotation of the Andromeda Galaxy* (1970) – Observational evidence of flat galactic rotation curves, leading to the dark matter hypothesis.
- **Perlmutter, S. et al.** *Measurements of Omega and Lambda from 42 High-Redshift Supernovae* (1999) – Discovery of the accelerating expansion of the universe, leading to the concept of dark energy.

4. Philosophical & Theoretical Inspirations

- *Bhagavad Gita* – Ancient discussions on universal intelligence, energy, and cosmic order.
- *Upanishads* – Early interpretations of reality, energy, and interconnected existence.
- *Vedas* – Ancient texts suggesting structured cosmic order and natural equilibrium.
- *Dnyaneshwari* (by Sant Dnyaneshwar) – Philosophical discussions on the interplay between the individual and the universal force.

5. Additional Scientific & Modern Theoretical Works Referenced

- Various NASA and ESA mission reports on gravitational anomalies (Pioneer anomaly, flyby anomalies).
- Research papers on quantum field theory and vacuum energy fluctuations.
- Studies on artificial gravity and vacuum energy engineering concepts.

Final Thoughts on Sources & Future Research

The ideas presented in *The Blueprint of the Universe* build upon a rich history of scientific discovery and philosophy while challenging conventional views to propose a structured energy equilibrium as the true nature of gravity, motion, and the universe itself.

This work is **not the conclusion, but the beginning of a larger scientific dialogue**—one that will require further research, experimental validation, and open discussion across physics, cosmology, and beyond.

For future researchers and thinkers: Let this bibliography serve as both a foundation and an invitation to continue expanding the frontiers of human knowledge.

Appendix

This appendix includes additional calculations, derivations, and technical proofs that support key arguments presented in The Blueprint of the Universe. It also contains an extended discussion on mathematical formulations, theoretical implications, and a glossary of essential terms to aid readers unfamiliar with technical jargon.

1. Additional Calculations & Mathematical Proofs

This section provides **detailed mathematical derivations** of the equations and principles introduced in the book, including:

1.1 Universal Energy Equation (Structured Vacuum Energy Framework)

Derivation of:

$$E = \int \rho_E dV$$

How **structured energy density interacts with mass to create gravitational effects.**

1.2 Chavan's Law of Gravity – Gravity as an Emergent Effect

Mathematical proof that gravity is **not an intrinsic force but a function of structured vacuum energy density gradients.**

Derivation of:

$$F_g = \nabla E_v = \frac{E_0 G M}{c^2 r^2}$$

Explanation of **why this replaces Newton's and Einstein's gravity models.**

1.3 Energy-Dependent Weight Equation (Testing Structured Vacuum Energy Effects on Mass)

Derivation of:

$$W = mg + \nabla E_v$$

Theoretical implications for **weight variations in different gravitational environments.**

1.4 Modified Kepler's Laws for Orbital Motion in UED

Modification of Kepler's third law based on structured vacuum energy interactions:

$$T^2 \propto \frac{1}{\nabla E_v}$$

Why **elliptical orbits emerge naturally from vacuum energy structuring.**

1.5 Black Holes as Structured Energy Sinks

Proof that black holes are **not singularities but structured energy zones maintaining cosmic equilibrium.**

Discussion of energy transitions near the event horizon using:

$$E_v(r) = E_0 \left(1 - \frac{GM}{c^2 r}\right)$$

2. Theoretical Implications & Extended Discussion

- How structured vacuum energy interacts with planetary motion, cosmic expansion, and artificial gravity.
- Why dark matter may not exist and galactic motion is naturally explained through structured energy interactions.
- Why hydrogen is the first and continuously forming energy-matter interface in the universe.
- Predictions on how experiments (Casimir effect variations, laser interferometry, gravitational lensing) could confirm the presence of structured vacuum energy fields.

3. Glossary of Technical Terms

To assist readers unfamiliar with technical jargon, the following glossary provides definitions of key concepts used throughout the book.

A. Universal Energy Dynamics (UED) Concepts

- **Structured Vacuum Energy** – The theory that vacuum is not empty but contains structured energy fields that influence gravity and motion.
- **Energy Equilibrium** – The principle that all motion and mass interactions follow an energy-balancing mechanism, rather than force-based laws.
- **First Disturbance** – The moment when structured energy first transitioned into matter formation, leading to the creation of hydrogen and early cosmic structures.

B. Chavan's Energy-Based Relativity (EBR) Terms

- **Energy-Based Relativity** – The idea that gravity and motion emerge from vacuum energy interactions rather than space-time curvature.
- **Modified Kepler's Laws** – Orbital motion predictions based on energy structuring instead of purely mass-based gravitational forces.
- **Weight Variations in Different Energy Fields** – How structured vacuum energy influences mass differently in different environments, leading to slight weight fluctuations.

C. Chavan's Law of Gravity Terms

Gravity as an Emergent Effect – The theory that gravity is not a fundamental force but a byproduct of energy density structuring.
Black Holes as Structured Energy Zones – The idea that black holes regulate cosmic energy rather than being infinite singularities.
Artificial Gravity Engineering – Using controlled vacuum energy structuring to generate gravitational effects in space.

Final Thoughts on the Appendix

This section serves as an additional resource for readers who want a deeper mathematical and conceptual understanding of the theories presented. The calculations and discussions here provide a solid foundation for future experimental validation and theoretical refinement of Universal Energy Dynamics (UED) and Chavan's Law of Gravity.

For researchers, physicists, and thinkers: May this appendix serve as a bridge between established physics and the next great scientific breakthrough.

Further Reading & Suggested Research

This section provides recommended books, papers, and research directions for those interested in expanding their understanding of physics, cosmology, energy dynamics, and the emerging framework of Universal Energy Dynamics (UED), Energy-Based Relativity (EBR), and Chavan's Law of Gravity.

1. Recommended Books & Papers on Fundamental Physics & Cosmology

These works have contributed to shaping the foundation of **modern physics, relativity, quantum mechanics, and cosmology.**

A. Classical & Modern Physics

- **Isaac Newton** – *Philosophiæ Naturalis Principia Mathematica* (1687) – The foundation of classical mechanics and gravity.
- **Albert Einstein** – *The Meaning of Relativity* (1922) – Explanation of space-time and gravitational curvature.
- **Richard Feynman** – *QED: The Strange Theory of Light and Matter* (1985) – Introduction to quantum electrodynamics.
- **Stephen Hawking** – *A Brief History of Time* (1988) – Overview of black holes, relativity, and the nature of time.
- **Max Planck** – *The Origin and Development of the Quantum Theory* (1920) – The early understanding of quantized energy.
- **Satyendra Nath Bose & Albert Einstein** – *Bose-Einstein Condensate Papers* (1924) – Studies on structured quantum states of matter.

B. Works on Energy & Alternative Gravity Theories

- **Nikola Tesla** – *Experiments with Alternate Currents of High Potential and High Frequency* (1891) – Early exploration of energy transmission.
- **Hendrik Casimir** – *On the Attraction Between Two Perfectly Conducting Plates* (1948) – Study of vacuum energy fluctuations.
- **Eric Verlinde** – *Emergent Gravity and the Dark Universe* (2016) – Proposing gravity as an

emergent phenomenon, an idea related to UED.

- **Vera Rubin & Kent Ford** – *Rotation of the Andromeda Galaxy* (1970) – Early evidence of missing mass (dark matter problem).

2. Suggested Research Directions for Future Scientific Exploration

While The Blueprint of the Universe introduces a new framework for understanding gravity, motion, and energy interactions, there are several areas that require further exploration, experimentation, and validation. Below are key scientific directions for future researchers and physicists to pursue.

A. Experimental Validation of Universal Energy Dynamics (UED)

- **Space-Based Casimir Effect Variations** – Testing how vacuum energy fluctuations change in different gravitational environments.
- **Laser Interferometry in Space** – Measuring potential light speed variations in structured vacuum energy fields.
- **Gravitational Anomaly Studies** – Revisiting space mission anomalies (Pioneer, flybys) to test for structured vacuum energy interactions.

B. Testing Chavan's Law of Gravity in Different Environments

- **Weight Measurements in Varying Gravitational Potentials** – Checking for vacuum energy-dependent weight fluctuations.
- **Pendulum Motion Variations** – Analyzing whether pendulum oscillations deviate due to structured vacuum energy gradients.
- **Hydrogen Regeneration Studies** – Investigating whether hydrogen formation is continuous in cosmic and lab environments.

C. The Role of Hydrogen in Energy Structuring

- **Testing Hydrogen's Interaction with Vacuum Energy Fields** – Observing whether hydrogen behaves differently in structured energy conditions.
- **Studying Hydrogen Isotope Distributions in Space** – Checking if their ratios correlate with structured vacuum energy predictions.
- **Hydrogen's Role in Biological Energy Transfer** – Investigating if structured energy equilibrium principles govern ATP synthesis and photosynthesis.

D. Artificial Gravity & Energy Field Manipulation

- **Vacuum Energy Engineering for Gravity Simulation** – Developing systems to generate artificial gravity using structured vacuum energy fields.
- **Quantum Coherence & Energy Optimization** – Studying whether quantum coherence in biological systems reflects structured energy dynamics.
- **Energy-Based Propulsion Concepts** – Exploring how spacecraft motion can be optimized by interacting with structured vacuum energy fields.

3. Final Thoughts: The Future of Research in Energy-Based Physics

- The universe is not just governed by force-based mechanics—it operates through structured energy interactions.
- Theoretical predictions must now be validated with real-world experiments, space missions, and laboratory tests.
- If UED, EBR, and Chavan's Law of Gravity hold true, they will revolutionize our understanding of the universe, from cosmology to space travel and artificial gravity.

For future researchers, physicists, and thinkers: The journey of discovery does not end here. This book is an invitation to explore the uncharted territories of energy-based physics. The next revolution in science awaits.

About the Author

Sandeep Chavan – Engineer, Thinker, and Architect of a New Scientific Paradigm

Sandeep Chavan is a mechanical engineering graduate from Pune University (2002) with extensive experience in industrial engineering, manufacturing, and planning. For the past 15 years, he has been dedicated to teaching physics, guiding students in understanding the fundamental principles of science and technology.

Beyond traditional engineering and education, Chavan has always been driven by a deep curiosity about the nature of the universe, the interplay of energy, and the limitations of existing scientific models. This pursuit has led him to develop groundbreaking theories that challenge Newtonian mechanics, Einsteinian relativity, and mainstream gravitational physics.

Vision & Motivation behind *The Blueprint of the Universe*

From the earliest scientific discoveries to modern physics, humanity has attempted to explain the fundamental forces governing reality. Yet, many unanswered questions remain:

- Is gravity truly a force, or is it an emergent effect of structured energy interactions?
  Why do galaxies rotate without the need for dark matter?
- Does space contain a hidden energy medium that dictates motion and cosmic equilibrium?
- Can we redefine relativity in terms of structured vacuum energy instead of space-time curvature?

In The Blueprint of the Universe, Chavan introduces Universal Energy Dynamics (UED), Energy-Based Relativity (EBR), and Chavan's Law of Gravity—a revolutionary framework proposing that the universe does not function through force-based mechanics alone, but through structured energy equilibrium.

His work is deeply inspired by Einstein, Newton, Tesla, Satyendra Nath Bose, and the foundational thinkers of both classical and quantum physics. Additionally, ancient Indian texts such as the Bhagavad Gita, Upanishads, and Vedas have provided philosophical insights into the interplay between universal intelligence and structured energy fields.

Contributions to Science & Philosophy

- **Universal Energy Dynamics (UED)** – A framework proposing that energy structuring governs gravity, motion, and cosmic interactions.
- **Chavan's Energy-Based Relativity (EBR)** – A post-Einsteinian model where gravity is **not space-time curvature but an effect of structured vacuum energy.**
- **Chavan's Law of Gravity** – A new approach that **eliminates the need for dark matter and redefines gravity as an emergent effect.**
- **Hydrogen as the First Structured Energy Form** – Proposing that **hydrogen is continuously forming as part of universal energy equilibrium.**
- **Artificial Gravity & Energy Engineering** – Investigating how **vacuum energy structuring could be used for space travel and gravity simulation.**

Chavan's work is not limited to **scientific theory—it is a call for new experiments, deeper exploration, and a fundamental rethinking of physics.**

A Message to Readers, Thinkers, and Scientists

The Blueprint of the Universe is **not just another book on physics—it is an invitation to rethink the very foundation of scientific understanding.**

- If Newton redefined motion and Einstein redefined relativity, this work aims to redefine energy and gravity as structured interactions rather than fundamental forces.
- If science is truly about discovery, then we must dare to challenge what is already known.
- This book is not the conclusion, but the beginning of a scientific revolution—one that will be shaped by future researchers, thinkers, and pioneers.

For those who question, for those who seek, and for those who believe that science is still evolving—this book is for you.The future of physics is waiting. Let's explore it together.

Call to Action / Open Invitation for Discussion

A New Scientific Dialogue Begins – Your Participation is Key

The Blueprint of the Universe is not just a book—it is a scientific proposition, a challenge to existing models, and an invitation to rethink our fundamental understanding of physics.

While this work lays down the foundation for Universal Energy Dynamics (UED), Energy-Based Relativity (EBR), and Chavan's Law of Gravity, true scientific progress requires open discussion, experimental validation, and continuous refinement.

This is a call to action for physicists, astrophysicists, engineers, researchers, and thinkers to join this effort in understanding the structured energy interactions that govern the universe.

1. Inviting Physicists, Scientists & Thinkers to Engage with This Theory

- If you are a physicist, cosmologist, or researcher in gravitational physics, quantum mechanics, or astrophysics, we invite you to analyze these ideas critically.
- If you are an experimental scientist, we encourage you to propose ways to validate structured vacuum energy effects.
- If you are a philosopher or deep thinker, we welcome discussions on the implications of structured energy equilibrium on existence, consciousness, and universal intelligence.

Your insights, critiques, and contributions will be invaluable in refining and expanding this framework.

2. Suggested Areas for Experimental Validation

To ensure that **these ideas move beyond theoretical discussion and into measurable science**, here are some proposed experiments:

- **Casimir Effect Variations in Space** – Measuring changes in vacuum energy interactions at different gravitational potentials.
- **Laser Interferometry for Light Speed Variations** – Testing whether vacuum energy structuring influences light propagation.
- **Hydrogen Regeneration Studies** – Investigating whether hydrogen is continuously forming as an energy-matter interchange process.
- **Weight & Pendulum Motion Tests** – Checking for small, structured vacuum energy

effects in different gravitational environments.
- **Artificial Gravity Engineering** – Developing controlled energy structuring to simulate gravitational effects.

Researchers are encouraged to explore these tests and refine methodologies for validating structured vacuum energy interactions.

3. Encouraging Scientific & Philosophical Dialogue

- Science progresses through questioning. The best way to validate or refine a theory is through open discussion.
- Philosophers and thinkers are invited to explore the deeper implications of structured energy equilibrium on reality, intelligence, and the interconnectedness of the universe.
- Academics, educators, and students are encouraged to engage in thought experiments, discussions, and theoretical expansions based on the ideas presented.

This book is just the beginning—the conversation must continue.

4. How to Get Involved

For Scientists & Researchers: Propose experimental tests, analyze the equations, and provide feedback. Conduct independent studies to explore the implications of vacuum energy structuring.

For Educators & Students: Engage in classroom discussions on how gravity, motion, and energy interactions could be reinterpreted. Use this work as a platform for academic debate and further exploration.

For Philosophers & Thinkers: Discuss the implications of structured energy equilibrium beyond physics—how it might relate to consciousness, existence, and the nature of intelligence. Bridge the gap between scientific understanding and philosophical inquiry.

Final Thought: The Future of Science is Open Inquiry

This is **not a closed book—it is an open invitation.**

- If **this theory is correct, it will redefine physics.**
- If **it is incomplete, it will lead to new discoveries.**
- Either way, **progress happens only when we dare to challenge the known.**

To all physicists, researchers, and seekers of truth—let's not just study the universe. Let's redefine it.

Disclaimer

This book, **The Blueprint of the Universe**, presents a theoretical framework based on Universal Energy Dynamics (UED), Chavan's Energy-Based Relativity (EBR), and Chavan's Law of Gravity. The concepts, hypotheses, and formulations discussed herein aim to expand scientific understanding and propose new perspectives on physics, cosmology, and fundamental forces.

Not a Replacement for Established Science

The theories presented in this book are not intended to replace the well-established frameworks of Newtonian mechanics, General Relativity, or Quantum Mechanics, but rather to extend and refine them where existing models encounter inconsistencies or unexplained phenomena.

A Work in Progress

Scientific theories evolve through continuous testing, peer review, and experimental validation. The ideas proposed in this book should be approached as hypotheses open to further exploration, discussion, and empirical verification.

No Immediate Technological Claims

While the book discusses potential applications in artificial gravity, space propulsion, and energy-based life sciences, these remain theoretical projections that require rigorous scientific and engineering validation before practical implementation.

Encouraging Open Scientific Discourse

The purpose of this book is to stimulate curiosity, inspire research, and challenge conventional paradigms in the pursuit of deeper scientific truths. Readers are encouraged to engage critically, conduct independent studies, and contribute to the ongoing evolution of scientific thought.

Responsibility of the Reader

The interpretations, applications, or further derivations based on the ideas presented in this book are the responsibility of the reader. No claims are made regarding the immediate applicability or accuracy of the theoretical models in real-world experimental physics without proper scientific validation.

By reading this book, you acknowledge that its content is exploratory in nature and meant to contribute to scientific dialogue and future research, rather than serving as an absolute or final explanation of universal laws.

Science progresses not by accepting all that is known, but by questioning what remains unknown.

Sandeep Chavan

Don't miss out!

Visit the website below and you can sign up to receive emails whenever SANDEEP CHAVAN publishes a new book. There's no charge and no obligation.

https://books2read.com/r/B-A-EPGPC-AIEUF

BOOKS 2 READ

Connecting independent readers to independent writers.

Did you love *The Blueprint of the Universe*? Then you should read *Energy in Action: The Soul Demystified*[1] by SANDEEP CHAVAN!

[2]

Energy in Action: The Soul Demystified by **Sandeep Chavan** offers a fresh and transformative perspective on life and existence, redefining the soul as energy in action—a dynamic force confined to the body during life and seamlessly reintegrating with the universal flow upon death. With a unique blend of philosophy, science, and spirituality, this book explores life as a continuous energy loop, governed by natural laws and sustained through the interplay of consciousness, desires, and interactions.

The book begins by demystifying concepts of the soul and consciousness, peeling away layers of mysticism to reveal their roots in universal energy. It redefines life as an energy game, where desires fuel growth, pain teaches resilience, and connections sustain the rhythm of existence. Through practical examples, such as the kindness ripple effect, the interconnectedness of ecosystems, or the role of instincts in decision-making, Sandeep Chavan makes these abstract concepts relatable and applicable to everyday life.

Key themes include balancing instinct and intellect, understanding subconsciousness, and the profound clarity that acceptance brings when navigating life's transience. The book also reimagines the concept of God, shifting focus from divine absolutes to universal energy, inviting readers to

1. https://books2read.com/u/3JwxAQ

2. https://books2read.com/u/3JwxAQ

Also by SANDEEP CHAVAN

Echoes of War: The Russia-Ukraine Conflict and Its Global Ripples
Shattered Lives
The War Economy
Borders Redrawn
Arms and Allies
Collateral Damage
Justice in the Ashes

Somewhere in the Darkness of the Cosmos
The Blueprint of the Universe

The Eternal Sage Chronicles
The Sage and the Sovereign
The Family Pillar
The People's Flame
The Warrior's Soul

The Fundamental Theory of Life & Existence
Energy in Action: The Soul Demystified
The Active Mind
Energy in Connection
Death: The Disconnection

celebrate life as it is—a vibrant, interconnected system driven by action, interaction, and transformation.

Purpose is highlighted as a subjective construct, created through individual experiences and contributions rather than dictated by universal absolutes. This perspective encourages readers to embrace the mystery of existence, fostering curiosity, presence, and intentional living. With the help of real-world examples and practical tools, the book offers a grounded approach to aligning with the energy loop, finding harmony, and creating meaning in everyday actions and connections.

Energy in Action is a thought-provoking guide that challenges conventional views while remaining accessible to all readers. It emphasizes simplicity, interconnectedness, and the beauty of living authentically. By bridging ancient wisdom with modern understanding, the book serves as both a philosophical exploration and a practical roadmap for anyone seeking clarity, connection, and fulfillment.

If you're ready to redefine your understanding of life and participate fully in the dynamic energy loop, this book is your perfect companion.

Read more at https://www.sandeepjchavan.com/.

Ultra Modern Life Philosophy

The Duality of Laughter and Tears

Compare to Accept

Standalone

It's Not AI, It's AHI - Amplified Human Intelligence

The Decision Paradox

The Four Sapiens

Malicious Script of Indian Polity

The IIT Legacy & Global Impact

The IIT Legacy & Global Impact

You are just a Version of Your Original

Win the Game You Didn't Choose

Always Keep Your Bags Packed

Reveal to Shield - Learn the Game of Facades

Engage Beyond Elections

Culture, Identity & Change: The Evolution of Indian Society

The Triangular Dynamics

Never Trust Without Doubt

If You Can't Change the Country, Leave the Country

Live Fearless with Fear

India: The Investment Magnet

Realignment

Powering a Billion Dreams

Universal Skill

Extension: From Brush to Algorithm

Human Side of Einstein

The Future Knows No Lines

Let's Bring the Generations to the Table

Unified Energy Dynamics

A Beat That Never Dies

Strategic Horizons

Right From Scratch

www.ingramcontent.com/pod-product-compliance
Lightning Source LLC
LaVergne TN
LVHW080011210926
848481LV00011B/2988
9798230563099